Advances in Biology

nova
science publishers
www.novapublishers.com

Advances in Biology

Advances in Biology. Volume 8
Charles D. Grant (Editor)
ISBN: 979-8-89530-020-6
eBook ISBN: 979-8-89530-042-8

Advances in Biology. Volume 7
Charles D. Grant (Editor)
ISBN: 979-8-89113-827-8
eBook ISBN: 979-8-89113-901-5

Advances in Biology. Volume 6
Charles D. Grant (Editor)
ISBN: 979-8-89113-712-7
eBook ISBN: 979-8-89113-790-5

Advances in Biology. Volume 5
Charles D. Grant (Editor)
ISBN: 979-8-89113-441-6
eBook ISBN: 979-8-89113-475-1

Advances in Biology. Volume 4
Charles D. Grant (Editor)
ISBN: 979-8-88697-999-2
eBook ISBN: 979-8-89113-074-6

Advances in Biology. Volume 3
Charles D. Grant (Editor)
ISBN: 979-8-88697-779-0
eBook ISBN: 979-8-88697-846-9

More information about this series can be found at
https://novapublishers.com/product-category/series/advances-in-biology/

Charles D. Grant
Editor

Advances in Biology

Volume 9

nova
science publishers

www.novapublishers.com

DOI: https://doi.org/10.52305/JDWC2175

Library of Congress Cataloging-in-Publication Data

ISBN: 979-8-89530-574-4 (Hardcover)
ISBN: 979-8-89530-665-9 (eBook)
ISSN: 2833-1753

Published by Nova Science Publishers, Inc. † New York

Contents

Chapter 1	**Gas Chromatography–Mass Spectrometry Analysis of Carbohydrates and Linoleic Acid in Tomato Leaves Depending on Growing Conditions: An Experimental and Theoretical Study** Svetlana Motyleva, Bojidarka Ivanova, Maria Vinokur and Elia Yamileva	1
Chapter 2	**Advances in Mass Spectrometry-Based Proteomics: Workflow and Recent Biological Applications** Thi Phuong Truc Nguyen, Thi Quynh Nhu Nguyen, Rajat Ujjainiya, Shantanu Sengupta and Thi Thu Hoai Nguyen	53
Chapter 3	**A Morphological and Anatomical Evaluation of Common *Plantago* Taxa** Muzaffer Mükemre, Abdullah Dalar and Murat Ünal	97
Chapter 4	**Macrocyclic Ivermectin (IVM) as an Anti-Malarial Drug** Zohaib Saifi, Afreen Inam and Rashid Ali	121
Chapter 5	**Plant Kunitz-Type Protease Inhibitors: A Review** Rodrigo Michelini de Oliveira Thomasi, José Antônio da Silva and Ana Hortência Fonsêca Castro	151

Chapter 6 **The Short-Term Liquid Nitrogen Storage of *Pyrostegia venusta* Seeds: A Disruptive *Ex Situ* Conservation Strategy** 175
Mairon César Coimbra, Israel José Pereira Garcia, Hérica de Lima Santos and Ana Hortência Fonsêca Castro

Chapter 7 **The Overlap of Hypoparathyroidism and Ankylosing Spondylitis: A Clinical and Radiological Investigations** 195
Taieb Ach, Nejla Amri and Asma Ben Abdelkrim

Index 213

Chapter 1

Gas Chromatography–Mass Spectrometry Analysis of Carbohydrates and Linoleic Acid in Tomato Leaves Depending on Growing Conditions: An Experimental and Theoretical Study

Svetlana Motyleva[1,*]
Bojidarka Ivanova[2]
Maria Vinokur[1]
and Elia Yamileva[1]
[1]Strogoorganic Online Gardening School, Moscow, Russian Federation
[2]Lehrstuhl für Analytische Chemie, Institut für Umweltforschung, Fakultät für Chemie und Chemische Biologie, Universität Dortmund, Dortmund, Germany

Abstract

This chapter is devoted to the study of the metabolic profile of tomato leaves (*Solanum lycopersicum* L.) by gas chromatography-mass spectrometry. This method is widely used to analyze the composition of plant extracts. We studied the effect of growing conditions (with and without earthworms) on the synthesis of primary and secondary metabolites in the leaves of the tomato cultivar Mikhnevets. The profiles of alcohol-soluble extracts were analyzed. A total of 49 metabolites have been identified. It has been established that the presence or absence of earthworms in the soil during tomato cultivation leads to a change in the

[*] Corresponding Author's Email: motyleva_svetlana@mail.ru

In: Advances in Biology. Volume 9
Editor: Charles D. Grant
ISBN: 979-8-89530-574-4

content of metabolites. The main differences are found in the accumulation of eight carbohydrates and their derivatives (maltitol, 2-o-mannobiase, myo-inositol, 1,5-anhydroglucitol, arabinofuranose, methyl-α-D-glucofuranoside, methylgalactoside and allofuranose), as well as oxylipin, a polyunsaturated fatty linoleic acid (9,12-octadecanoic acid). The findings support the role of linoleic acid in signaling processes that activate protective genes in tomato leaves in response to growing conditions. The results of metabolite analysis in tomato leaves using not only gas chromatography-mass spectrometry are described in the chapter. We also use the innovative approach of stochastic dynamics, its model equations designed to expand the capabilities of analytical mass spectrometry in accurate quantitative and 3D molecular and electronic structural analysis. Theoretically and experimentally, the complexity of plant samples demonstrating matrix effects has been revealed. In plant extracts, analytes can not only be structurally very similar, but also produce mass spectra that are difficult to identify. The problem arises not only from the frequent lack of reference materials for many unknown metabolites, but also from the chemical indistinguishability of the metabolites from the chemicals used to prepare the extract. In addition, there may be incomplete chemical derivatization of metabolites, degradation processes, or competitive adverse reactions which may account for 5% to 15% of the spectral data. Such statistics support the view that the development of new theoretical approaches to the processing of mass spectrometry data and their additional use in metabolomics methods can significantly expand the capabilities of mass spectrometry to identify and annotate unknown analytes in complex samples. The equations of the mass spectrometry model of stochastic dynamics and the methods of metabolomics used by us as diverse approaches to the structural analysis of metabolites have shown that these independent methods complement each other. In this regard the methods of metabolomics and mass spectrometric stochastic dynamics are used in this chapter to determine the structural determination of linoleic acid, 1-methyl-α-D-D-glucofuranose, α-D-glucopyranose, α-D-arabinofuranose, β-D-galactofuranose, 1,5-lactone 1,5-anhydro-D-glucitol and 1-methyl-β-D-galactoside in methanol extracts. The last-mentioned analytes are particularly complex molecular objects requiring 3D mass spectrometry structural analysis, due to the lack of their regioselective derivatization, the presence of isomers, significant conformational flexibility, the possibility of intramolecular effects of proton and charge transfer, intramolecular rearrangement and cyclization reactions, respectively. The complementary methodological approaches of mass spectrometry, stochastic dynamics, and chemometry used by us as tools for the quantitative and structural determination of metabolites have made it possible to achieve excellent and accurate characteristics of the assigned metabolites, which showed |r| = 0.99987 (omics method) and |r| = 1 (stochastic dynamics mass spectrometry). Thus, new excellent characteristics of the method of the present study emphasize the prospects of the usage of the stochastic dynamics model equations in metabolomics as an innovative tool for determining the three-dimensional molecular structures of analytes.

Keywords: *Solánum lycopérsicum* L., leaves, gas chromatography and mass spectrometry, metabolomics, stochastic dynamics, quantum chemistry, structural analysis

Abbreviations and Acronyms

15ADG	1,5-Anhydroglucitol
1MADGF	1-Methyl-α-D-glucofuranose
ADMP	Atom-centered density matrix propagation
ANOVA	Analysis of variance (chemometric method)
BDGF	β-D-galactofuranose
BOMD	Born–Oppenheimer molecular dynamics
BSTFA	N,O-*bis*(trimethylsilyl) trifluoracetamide
CB	Carbohydrate
CID	Collision induced dissociation (mass spectrometric operation mode)
DFT	Density functional theory (quantum chemical method)
DoF	Number of points – number of parameters
D_{QC}	Quantum chemical diffusion parameter according to Arrhenius's theory
D_{SD}, D_{SD}''	Stochastic dynamic diffusion parameters according to our theory
Edel	D*eletion* energy (natural bond orbital analysis)
ECP	Effective core potential
ESI	Electrospray ionization (mass spectrometric method)
ESP	Electrostatic potentials
GC–MS	Gas chromatography–mass spectrometry
GS	Ground state
I	Intensity of mass spectrometric peak (variable)
LA	Linoleic acid
LMW	Low molecular weight (analyte)
m/z	Mass-to-charge (mass spectrometric variable)

MD	Molecular dynamics
MM	Molecular mechanics
MS	Mass spectrometry
MS/MS	Tandem mass spectrometric operation mode
MP2	Second-order Møller–Plesset perturbation theory
NBO	Natural Bond Orbital (quantum chemical method)
PCM	Polarizable continuum method
PES	Potential energy surface
$\|r\|$	Statistical coefficient of linear correlation (chemometrics)
RT	Retention time
sd(yEr±)	Standard déviation (chemometrics)
se(yEr±)	Standard error (chemometrics)
TS	Transition state

1. Introduction

Tomato (*Solanum lycopersicum* L.) is one of horticultural crops having high economic value. It is widely cultivated worldwide, using *ca.* 420 000 ha of agricultural land only in Europe [1]. The development of plants is influenced by genetic basis and soil conditions.

Despite numerous studies on the influence of soil composition on the growth processes and biochemical composition of tomato plants, the effect of the presence or absence of earthworms in the soil on the growth and development of tomato plants is still poorly understood.

When grown on substrates with earthworms, a decrease in the amount of certain metabolites was observed in tomato leaves [2–6] in comparison with a substrate without them. Primary metabolites — carbohydrates, lipids and secondary metabolites — organic acids, phenolic compounds are intermediate or end products of a complex network of biochemical pathways involved in plant metabolism [7–9]. Carbohydrates are essential compounds for all living organisms, regulating that regulate many vital biological processes [10]. Lipids — esters of polyunsaturated fatty acids — regulate mechanisms of plant development and their adaptation to environmental conditions [11]. Gas chromatography–mass spectrometry has proven to be an invaluable approach to determine the position of double bonds in linoleic acid [12].

Metabolite profiling is widely used to diagnose and understand the mechanisms of plant response and adaptation to a wide range of stresses or to changes in growing conditions [5–9].

In recent decades, analytical mass spectrometry has become indispensable for the quantitative and structural determination of analytes in complex biological [13], environmental and food samples [14-16]. It is considered to be the golden standard of analytical practice, as it provides highly reliable and accurate analytical data on metabolites [15, 17].

Carbohydrates have a diverse structure based on different types of building blocks (monosaccharides), ring size (pyranoses and furanoses), configuration (α and β), regiochemistry (different binding schemes), branching, chain length, repeating links, *etc*. Due to the complexity of their unique performance carbohydrate identification are still a big challenge [17].

Recently, many studies have focused on mass spectrometric approaches in the field of glycomics and glycoproteomics

Furthermore, methodological MS developments and applications to examine carbohydrates have resulted to systematic and in-depth knowledge of fragmentation processes in various ion activation MS methods. The fundamental carbohydrate chemically bond cleavage modes are commonly designed *via* the nomenclature developed by Domon–Costello [18], Stephens and coworkers [19], as well as Spina and coworkers [20], respectively.

There is a lack of consensus on a single instrumental method capable of identification, accurate quantitation, and 3D molecular structural determination of carbohydrates in a rapid manner, due to aforementioned significant structural diversity and chemical reactivity of these analytes [21-24].

Metabolomics has been, in fact, revolutionized by instrumentation newcomers in chromatography, ion mobility spectrometry, mass spectrometry, nuclear magnetic resonance, chemical crystallography, and other analytical methods and technologies. Thus, this method allows us not only to identify and annotate a large number of cellular metabolites, but also to evaluate changes in metabolic biochemical reactions and reactivity in real time.

The almost routine implementation of robust methods of structural analysis highlighted above into the field of metabolomics has spurred a development and concurrent proliferation of methods for data-processing of measurable variables of metabolites in addition to software tools for processing and analyzing metabolomics datasets of measurable variables. The same is true for elaboration theoretical methods and writing software for predicting chemical properties measured by the corresponding analytical instruments, particularly, underlying the methods of mass spectrometry [25]. In order to evaluate that experimental tasks and obtained raw MS measurands are appropriately designed, processed, and interpreted, there is chiefly implemented statistical and chemometrics quantitative criteria on reliability of both the experimental and theoretical metabolomics data.

Moreover, it has been found that the format of raw measurands and the type of analytical instrumentation have a significant impact on data analysis. The statistical assessment of metabolomics data appears to be a task of bioinformatics, among others [26].

Modern analytical equipment contains software for the calculation of results. However, it cannot always provide the calculation of the mass spectrometry analysis results based on 3D molecular and electron-structural analyses [27].

At the same time, it should be emphasized that the variations in oscillations of mass spectrometric measured quantities may depend not only on the elemental composition received as a result of (bio) chemical reactions, but also on the conditions of the experiment and the nature of the biological sample [15]. It can be, for instance, (i) the human factor associated with sampling, sample preparation, chemical synthesis of metabolite derivatives, etc.; (ii) the variation in the parameters and factors of the instrument, such as temperature, intra- and inter-laboratory variation in the characteristics of the instrument; (iii) inter-device variation; (iv) variation between ionization methods and analysis techniques; (v) unknown biological variation, such as the quantity of the cells tested, the body liquid concentration, etc. [27, 28]. In addition, the physicochemical properties of analytes can have a significant impact [29, 30].

The above-mentioned fact has created a unique requirement of the development of not only analytical mass spectrometric instrumentation, but also theoretical methods for data processing of measurands in order to fill in bioinformatics gaps of knowledge about metabolomics identification and annotation of low molecular weight chemical composition of biological, environmental, and food samples [31].

In the context of the preceding paragraphs, together with metabolomics methods this chapter uses the innovative stochastic dynamics of mass spectrometric model equations (1), (2), and (6) [32–37] as new theoretical method capable of determining highly precisely and reliable 3D molecular and electronic structures of metabolites. The application of equation (2) to determine 3D molecular and electronic structures of N-glycans has resulted in excellent method performances, showing $|r| = 0.9975_1$–0.9792 [35]. It produces exact ($|r| = 1$) quantitative and 3D structural analysis of metabolites in biological fluids such as urine or biota [33, 34], while equation (1) shows excellent-to-exact determination of native cyclodextrins examining low m/z range of analyte product ions. The application of equation (6) [36] to study metabolite in plants has resulted to $|r| = 0.9963_6$–0.9898 determining silylated 4- and 2-D-desoxy-glucose [37].

1.1. Stochastic Dynamics Mass Spectrometry

Stochastic dynamics mass spectrometry introduces model equations (1) and (2) [32–37], processing exactly measurands from MS experiments, where 'I' means absolute intensity data on observable parent or product ions of analytes. Since, the data-processing of measurable variables is performed per short spans of scan time there is exact processing of fluctuations of measurands using the formulas given below; furthermore, they are applicable to any span of scan time. Equations (1) and (2) allow us to determine the exact 3D molecular and electronic structures of molecules accounting for subtle electronic charge redistribution, tautomeric effects, protomers, and isotopomers of the examined analytes.

$$D_{SD}^{tot} = \sum_{i}^{n} D_{SD}^{i} = \sum_{i}^{n} 1.3194.10^{-17} \times A^{i} \times \frac{\overline{I_i^2} - \left(\overline{I_i}\right)^2}{\left(I_i - \overline{I_i}\right)^2} \quad (1)$$

$$D_{SD}^{'',tot} = \sum_{i}^{n} D_{SD}^{'',i} = \sum_{i}^{n} 2.6388.10^{-17} \times \left(\overline{I_i^2} - \left(\overline{I_i}\right)^2\right) \quad (2)$$

The stochastic dynamics mass spectrometric structural analysis is based on complementary employment in formulas (1) and (2) and Arrhenius's equation (3). Chemometrics assessment of the statistical significance of linear relation $D''_{SD} = f(D_{QC})$ is carried out.

$$D_{QC} = \frac{\prod_{i=1}^{3N} \nu_i^0}{\prod_{i=1}^{3N-1} \nu_i^s} \times e^{-\frac{\Delta H^{\#}}{R \times T}} \quad (3)$$

The method is developed to determine energetics of molecular species *via free Gibbs energy* parameter and Arrhenius's activation data on chemical reactions observed in MS conditions using connections among MS measurands, kinetics, molecular thermodynamics, and diffusion of molecular species. The transition state theory describes quantitatively aforementioned processes.

Chemometrics evaluate the statistical significance of relation $D''_{SD} = f(D_{QC})$; thus, to establish the link between MS datasets of measurands and analyte 3D molecular conformations and electronic structures.

Methodological developments of equations (1) and (2) have resulted in formulas (4) and (5). They determine exactly the temporal distribution of measurands depending on experimental conditions, parameters and factors.

$$D''_{SD,I} + D''_{SD,m} = \left|r_{I,m}\right| \times \sqrt{\overline{I^2_{I,q}} - \left(\overline{I_{I,q}}\right)^2} \times \sqrt{\overline{I^2_{m,q}} - \left(\overline{I_{m,q}}\right)^2} \quad (4)$$

$$D''^{,q}_{SD} \approx 2 \times \frac{A^q_D}{A^q_I} \times I^{TOT,q}_{av} \quad (5)$$

Equation (6) is derived, using formula (2). It predicts highly accurately mass spectra of molecules, depending on their 3D molecular and electronic structures. The equation applies approximations $<I^2> \, > \, <I>^2$ and D''_{SD}□D_{QC}. The I_{SD}^{Theor} is theoretical intensity data on product MS peaks of qth ion in certain experimental MS conditions. However, the formula (6) does not account for fluctuations of measurable variables.

$$I^{Theor}_{SD} \sim \left(2.6388.10^{-17} \times D_{QC}\right)^{1/2} \quad (6)$$

The formula (6) determines isomers of analytes, their, protonomers, and isotopomers when intensity data on product MS peaks are statistically significantly different. Equation (6) yields to following performances describing theoretically MS spectra of LMWs.

1.2. Quantum Chemical Data

Gaussian, Dalton, and Gamess-US [38–40] program packages were used to calculate quantum chemically molecular species. Outputs were visualized by

GausView03 [41]. DFT optimization was performed by means of *ab initio* Hartree–Fock, second- and third-order Møller–Plesset perturbation theory, as well as B3PW91 functional [42]. Truhlar's DFT functional M06-2X was used, as well [43, 44]. The application of MP2/cc-pVDZ level of theory in optimization of large molecular systems has shown good agreement of optimized molecular structural parameters and singly crystal X-ray data as well as DFT-optimized geometries with M06-2X or B3LYP-D3 functional [45].

Bernys's algorithm was utilized for determining ground state parameters. The stationary points on PES were obtained using standard harmonic vibration analysis. Minima of energy are confirmed, due to absence of imaginary frequencies of second-derivative matrix. Dunning's basis set cc-pVDZ and quasi-relativistic effective core pseudo potentials from Stuttgart–Dresden(–Bonn) (SDD and SDDAll) [http://www.cup.uni-muenchen.de/oc/zipse/los-alamos-national-laboratory-lanl-ecps.html] and effective core-potential basis LANL2DZ (Los Alamos National Laboratory 2 double ζ) were applied, as well. The application of LANL2DZ including relativistic effective core potentials for inner electrons is added with polarization functions used to C (d; 0.587), N (d; 0.736), and O (d; 0.961) atoms [46–48]. LANL2DZdp basis together with ECPs have been used to examine the molecular properties of silicon-containing species, as well [49].

The doubly polarized triple-ζ-basis sets added by diffuse functions aug-cc-pVDZ was used. Models of parent and product mass spectrometric ions in solution were examined using explicit super molecule and mixed-solvation approach to micro hydration *via* PCM methods. Equilibrium and transition states were confirmed by means of frequency analysis. The electronic structure analysis uses primarily two basic approaches: (i) Born–Oppenheimer approximation, which separates nuclear motion from electronic one, and (2) independent particle approximation, which presents total electronic wave-function description in form of Slater determinant or one electron wave-functions. There are static and dynamic electron correlations. An accurate and appropriate approach to solving multi-configurational problems in excited states of molecules is multi-reference CI and MCSCF/CASPT2 methods. Their practical applicability to excited state calculations is limited to small sized molecular systems. The application of *ab initio* MD approaches to excited state problems is a very challenging research task. The MD computations were carried out *via ab initio* and DFT BOMD and ADMP methods. The MD computations were carried out by means of M062X functional. The trajectories were integrated by means of Hessian-based predictor-corrector approach with Hessian updating for each step on BO PES. The step sizes were 0.3 and 0.25 amu$^{1/2}$Bohr. The trajectory analysis stops when: *(a)* Centres of mass of a dissociating fragment are different at 15 Bohr, or *(b)* when number of steps exceed given to as input parameter maximal number of points. The total energy was conserved during computations within at least 0.1 kcalmol^{-1}. The computations were carried out by means of fixed trajectory time speed ($t = 0.025$ fs) starting from initial velocities. The velocity Verlet and Bulirsch–Stoer integration methods were applied.

Allinger's molecular mechanics force field MM2 was used [50, 51]. The low order torsion terms are accounted with higher priority rather than Van der Waals interactions.

Important approaches to study excited state time-dependent isomerization reactions of radical-cations of aromatic heterocyclics systems are classified as adiabatic and diabatic ones. In cases of adiabatic MD computations, the molecular object is restricted to one electronic state during the trajectory analysis. Conversely, in cases of diabatic MD computations, there are jumps of the molecular object from one PES to another one~~,~~ using suitable algorithms. The process of ion-radical formation, due to high energy irradiation, is regarded as primary step in DNA damage within the framework of direct mechanisms. Due to these reasons this chapter studies radical-cations of the entitled analytes using both adiabatic and non-adiabatic methods. The obtained energetics of the species is corrected by basis set superposition error methods [52, 53].

Adiabatic *ab initio* and DFT Born–Oppenheimer MD approach has been detailed [54], while among nonadiabatic MS the theory of surface hopping dynamics [55] time dependent on DFT methods with long-range corrected density functional and Tamm–Dancoff approximation [56] is used. The variations of the surface-hopping approach have been discussed [57] and the details of the ~~on~~ time dependent DFT method have been described, as well [46–48, 58, 59].

As quality criterion determining the minimum number of QM/MM computations of molecular energies, number three is used according to data on work [60].

The discussed data are on high accuracy CBS-QB3, M06-2X/Jun-cc-pVTZ, and M06-2X/6-311+g(2df,2p) approaches and results from work [61]. The k(E) within the framework of Rice–Ramsperger–Kassel–Marcus theory is given by equation (10) [62]. The chemometric parameters presented, so far, range from excellent to accurate among the theoretical levels of computation, including the DFT and CBS-QB3 methods, highlight the advantages of DFT methods in balancing accuracy and computational cost. Therefore, despite the fact that there is a method chiefly applied in M062X DFT, the calculated D_{QC} parameters of equation (3), respectively, I^{Theor}_{SD} data on equation (6), herein, are highly accurate and reliable ones. The theoretical parameters reported to this chapter correlate to the above given statement, as well.

The electron density distribution of the ionic species of metabolites provides crucial information about the chemical reactivity, molecular properties, and interaction modes of the species. The physical parameter which is frequently used in the given context is molecular electrostatic potential [63]. It can be obtained highly accurately *via ab initio* and via DFT quantum-chemical total energy computations [53]. Furthermore, experimentally, the above-mentioned parameter can be obtained by means of high-resolution spectroscopic approaches to correlate the theoretical and experimental molecular properties of the analytes. ESP reflects the energy of intermolecular/ionic interactions, since the energetics of the molecular species is determined in the terms of electrostatics, polarization, exchange, and more energy terms, however, the former is dominant. Despite, the fact that the chapter herein provides ESP parameters, obtained at high level of theory, in balancing between computational costs and accuracy among theoretical approaches used to compute properties of large molecules there has been shown that *ab initio* HF/STO-3G method yields to appropriate results in 163 seconds computing oligonucleotides [63].

The natural bond orbital analysis of silicon species was performed, as well [65, 66]. The natural iconicity parameter (i_{AB}) is calculated using equation 4.18 (see p. 63 [65].)

The computations of silicon-containing species accounting for scalar and spin-orbit relativistic effects, in addition to total relativistic corrections, were carried out. Molecular ions containing heavy atoms exhibited relativistic effects. They play an important role in the reliable description of electronic effects and the structure of molecular ions and their chemical bonding properties. The relativistic effect in ^{29}Si-containing ions cannot be regarded as a spin-orbit coupling one. It is described as a result of scalar relativistic contraction of electronic s-shells [67, 68]. Among relativistic methods for computing properties of heavy-atom-containing molecular ions, the second-order Douglas–Kroll–Hess method is used [69]. Spin-orbit computations were performed using ECP spin-orbit approach to silicon containing cations. Silicon-cation has been tested in silicon-containing molecules using the basic Slater-type feature sets used to represent valence pseudoorbitals. Thus, the contribution of silicon cations to the energy of molecular ions is taken into account [70]. Herein, CAS(4,4)/6-31G** level of theory is used. Among the theoretical methods a correlative analysis of energetic, and vibration frequencies, and I^{Theor}_{SD} parameters of equation (6) is also carried out.

The theoretical *ab initio* and DFT methods were used to determine the energetic, molecular and electronic structures, and vibration spectroscopic properties of molecules in GSs and TSs. Their accuracy depends on the level of theory. Depending on molecular size and presence of heavy metal atoms, even *chemical accuracy* can be achieved, which denotes cases, when theoretical thermochemical parameters of LMWs and experimental data should be different in ~1 kcal.mol^{-1} [71].

However, as the size of the molecule increases there are several approximations in theory that are used to balance accuracy and computational costs. The so-called *gold standard* of theoretical methods combined with the cluster of single, double, and perturbative triple excitations (CCSD(T)) of the theory level is inferior to thermochemistry parameters with chemical accuracy. However, medium-sized molecular systems or LMWs containing a large number of heavy atoms or ions can be calculated with high accuracy, balancing between accuracy and cost of their analysis. Computationally cheaper tools for the above purposes are DFT methods.

However, it is not only the theoretical method, but also the basis set size that affects the computational cost, accuracy, and reliability of the data. Common basis sets are Slater- and Gaussian-type; e.g., Dunning base set families that are consistent with correlation, Jensen pc-n basis sets, and much more [72]. Despite the well-known problem of estimating multicenter two-electron integrals of Slater-type basis sets in comparison with GTO, they are physically more valid for molecules, since the exact wave function behaves in infinity according to the same exponential dependence. The basic set of STO-3G is a cost-effective approach [73] and is valuable, among other things, for the development of new algorithms [74]. Modifications of STO-3G basis take into account electronic and magnetic properties and correlation [75]. Good performance has been reported to correlate between theoretical and experimental ^{1}H nuclear magnetic resonance shieldings. State-specific approximations of PES *via* STO-3G are shown (2022) [76]. The results of HF/STO-3G and HF/6-31G** computations of LMWs showed comparable general trends in the study of geometry parameters.

However, the first method reduces the number of computational cycles required to converge to the optimized molecular geometry, while reducing the overall computational time [77]. For the calculation of high-molecular analytes, a small basic set of STO-3G and improved modifications of SIT are used [78–82].

However, when calculating large molecular systems, an unexpected proton transfer effect often occurs when optimized with the SRT-3G base set [83]. In such cases, if any, calculations of the control molecular structures devoid of proton transfer effects in STO-3G optimization and parallel calculations at a higher level of theory should be performed. In our cases, complementary SRT-3G and the above-mentioned high-precision approaches are used to optimize the molecular structures of species.

Although accuracy of Slater-type basis sets is lower, they are about ten to 50 times faster compared to the high-precision G4 method [84]. The maximum error in calculating reaction enthalpies in unit $kcal.mol^{-1}$ of HF/STO-3G method is 13.5 $kcal.mol^{-1}$ [84] or 0.0215 a.u. A scale factor of 0.67 is used. The scaled theoretical frequency of tension oscillations of SRT-3G is in good agreement with experimental data [85, 86].

Thus, the literature studies have revealed a contribution to the analysis of carbohydrates and linoleic acid in tomato leaves by the GC–MS method, including the use of stochastic dynamics method for in-depth molecular structural analysis of plant mass spectrometric massive of spectra.

2. Results

2.1. GC–MS Leaf *Solanum lycopersicum* L. Analysis

To evaluate differences in the metabolite composition of tomato leaves grown in the soil with the presence or absence of earthworms a non-targeted analysis was carried out, using GC–MS JMS-Q1050GC chromatograph coupled to mass spectrometry. JMS-Q1050GC is equipped with hyperbolic quadrupole of the largest size in its class; it has a large ion inventory, resulting in a wide dynamic range of ion counts. Its capability allows measuring a sample without the condition changes depending on analyte concentration of the sample.

In addition, hyperbolic shape quadrupole shows excellent ion permeability, realizing enhanced sensitivity, especially in the high mass region. By controlling experimental factors, such as the temperature of the quadrupole unit with high accuracy, the stability of m/z axis is greatly improved. NovaSpec software combined with a library search algorithm was used to identify and annotate compounds in tomato leaf extracts. The tool allows automatic detection of analytes based on the presence of accurate mass of precursor ions with a mass tolerance of 5 parts *per* million. To reduce chemical interference from the sample matrix, the following filters were set: a mass tolerance for alignment of 5 ppm, an intensity tolerance of 30%, a total intensity threshold of 1×10^5, a maximum shift of 0.5 min, and a signal-to-noise ratio threshold of 10.

Next, the datasets were manually processed for identifying chemical markers. The analyte identification was based on the detection of at least three diagnostic ions — one precursor ion and two fragment ions — with a mass accuracy of < 5 ppm. For this purpose, the NIST 15 program was additionally used. Forty-nine metabolites were reliably identified via metabolomics methods (Figures 1 and 2).

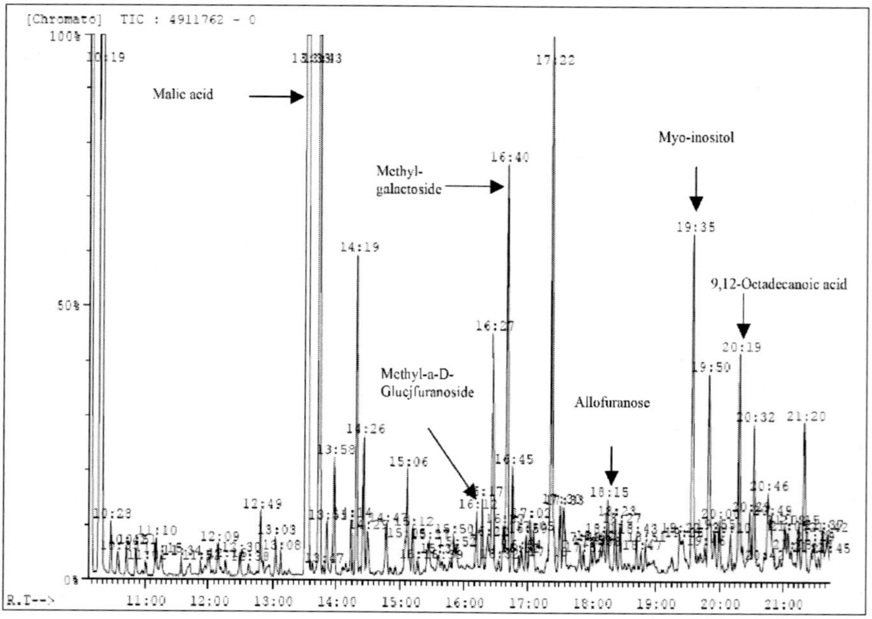

Figure 1. Chromatogram of tomato leaves (cv. Mikhnevets); RT = 11–21 mins; standard terpene mixture and *S. galapagense* (LA1137) at concentration 10 μg.mL^{-1} (see also Figure 2).

There are 29 compounds identified as CBs and their derivatives (Figure 3). Major differences were found in the accumulation of eight CBs and their derivatives; e.g., maltitol, 2-α-mannobiose, myo-inositol, 1,5-

anhydroglucitol, arabinofuranose, methyl-α-D-glucofuranoside, methyl galactoside, and allofuranose; as well as linoleic acid.

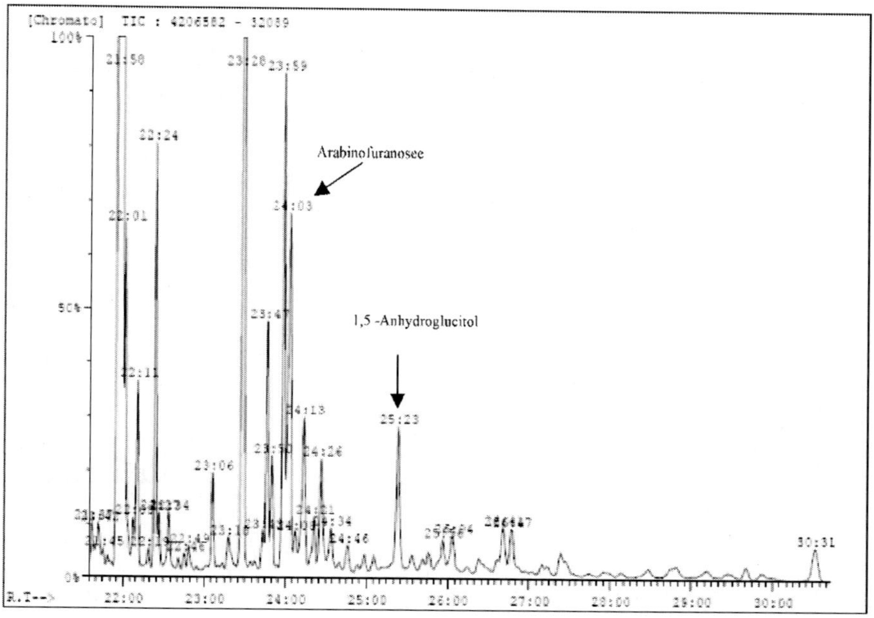

Figure 2. Chromatogram of tomato leaves (cv. Mikhnevets); RT = 22–30 mins.

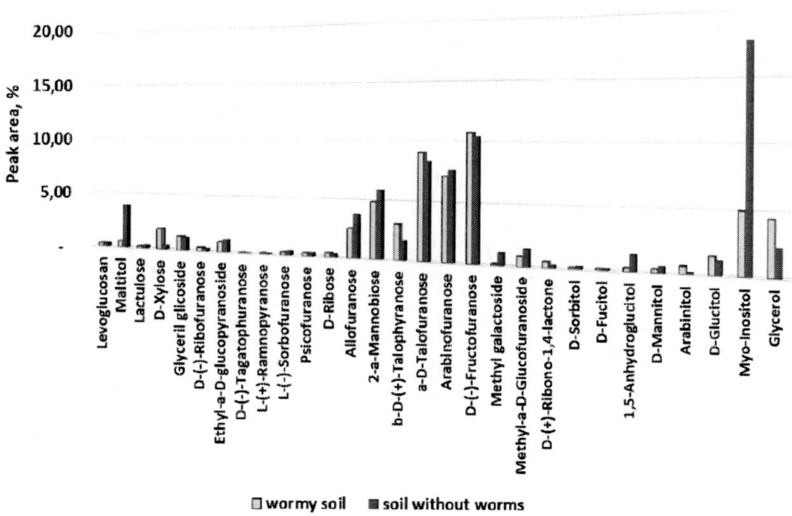

Figure 3. Comparative carbohydrate composition in tomato leaves (cv. Mikhnevets).

	substance	wormy soil	soil without worms
1	L-Alanine	0.27	0.15
2	L-Valine	0.21	0.11
3	d-Proline	0.26	0.24
4	Tyramine	0.25	0.30
5	9.12-Octadecanoic acid	0.50	0.85
6	Palmytic acid	1.19	1.04
7	Sebacic acid	0.43	0.32
8	Phytol	0.20	0.19
9	2.3.4-Trihydroxybutiric acid	0.11	0.11
10	Gluconic acid	1.20	1.64
11	Buthylboronic acid	0.08	0,10
12	2-Butenedioic acid	0.53	0.22
13	Glyceric acid	0.11	0.08
14	Oxalic acid	0.15	0.07
15	2.3.4.5-Tetrahydroxypentanoic acid	0.07	0.07
16	Pyruvic acid	0.26	0.43
17	2-Oleoylglucoside	1.33	1.39
18	Malic acid	11.79	11.97
19	Hexonic acid	0.04	0.04
20	Pyrrolidine-2-carboxylic acid	0.03	0.05

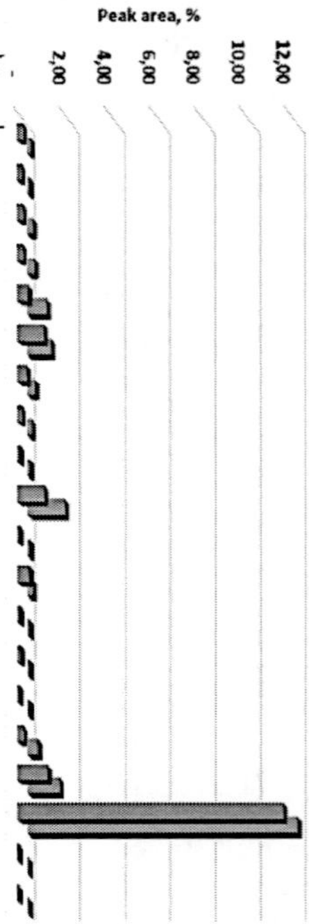

Figure 4. Comparative composition of amino acids, organic and fatty acids in tomato leaves (cv. Mikhnevets).

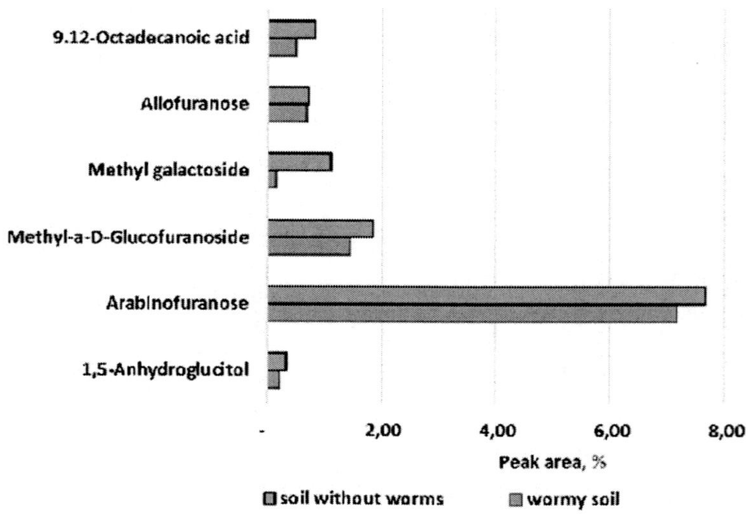

Figure 5. Comparative composition of carbohydrates and linoleic acids in tomato leaves (cv. Mikhnevets).

Twenty metabolites were identified as amino acids, fatty acids, and organic acids (Figure 4).

Further, the theoretical analysis of six compounds was performed, *i.e.,* 1,5-anhydroglucitol, arabinofuranose, methyl-α-D-glucofuranoside, methyl galactoside, allofuranose, and linoleic acid, respectively (Figure 5).

2.2. Stochastic Dynamics Mass Spectrometric Structural Analysis of Metabolites

In the aspect of multiple omics-methods for identification and annotation of plant metabolites, this sub-section of the chapter identifies and structurally determines analytes on the base of the described above stochastic dynamics approach and model equations (2), (3), and (6), respectively.

Although the debates about the methodological characteristics of omics approaches cannot be questioned, in general, questions regarding the reliability of the annotation of unknown analytes based on metabolomics often arise that should be addressed. Due to the complexity of the 3D molecular conformation and electronic effects of even LMWs, as well as the ability of analytes to stabilize a set of tautomers and protomers, as well as various isotopomers, metabolomic approaches are often inferior to comparable method characteristics. In such cases, if available, reliable identification and annotation of metabolites require the use of various theoretical methods in order to improve the effectiveness of the method. The burden of proof in such cases should not only increase the efficiency of the method, but also increase the reliability of analytical information about the molecular structure of the analytes under study.

Since the metabolomics method does not employ a theory such as stochastic dynamics, the mass spectrometry approach presented here and its model equations (1) and (2) can be considered as a methodological development of mass spectrometry methods for quantitative and three-dimensional structural analysis. In this context, innovation equations (1), (2) and (6) do not claim to be a new omics method that produces a data processing algorithm for decision-making on measurable variables of MS metabolites. Rather, they deal with the assessment by means of quantitative chemometric criteria of the statistical correlative significance of the measured variable intensity of the analyte MS peak and the corresponding 3D molecular and electronic structure of the substance producing the mass spectrometry observed peak. The described task largely involves the evaluation of possible tautomeric forms and protomers of parent and product ions of analytes; their ability to carry charge intra- and intermolecularly; the effects of intramolecular rearrangement and proton transfer; diversity of molecular isotopologies; the possibility of donating and accepting protons, and much more.

Thus, the question of whether the molecular model of parent and product ions is valid raises many basic questions related to the chemical reactivity, thermodynamic stability, and electronic structure of the analytes.

As already noted, because the mass spectrometry theory of stochastic dynamics and its model equations arose in the development of precise methods of analytical quantitative and 3D structural analysis, this subsection attempts to answer the question of why the formulas under discussion should presumably be applied to metabolomics methods for identification and annotation?

Thus, Figure 6 shows the GC–MS spectrum of a standard sample of silylated linoleic acid together with its MS fragmentation scheme, while Figure 7 shows the correlation between MS measurable variables of metabolite and NIST data on standard according to DB1 together with the results from the reference [41] studying silylated linoleic acid, as well.

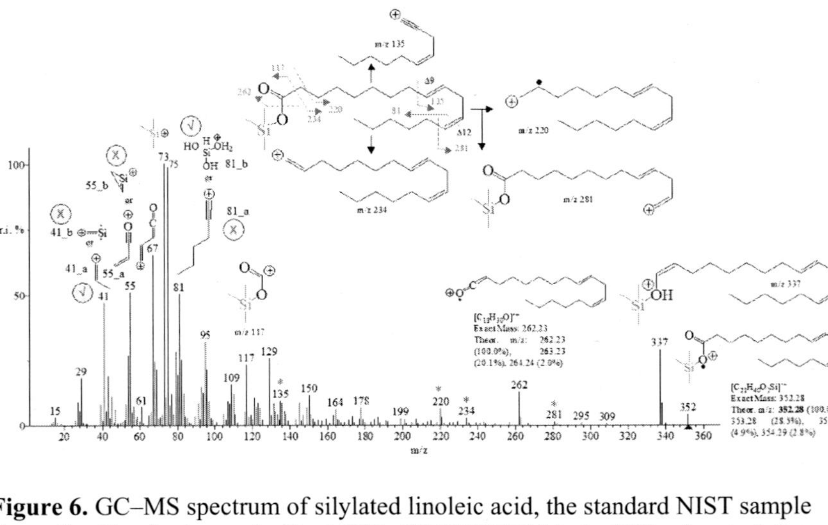

Figure 6. GC–MS spectrum of silylated linoleic acid, the standard NIST sample [https://webbook.nist.gov/cgi/inchi?ID=C56259075&Mask=200]; chemical diagrams of mass spectrometrically observed parent and product ions and cation-radicals of the analyte.

As can be seen, there is a wide correspondence between the variables measured which results in excellent method performances ($|r| = 0.9831_4$) when evaluating 16 product ions of the analyte. The exact parameter $|r| = 1$ was obtained by studying the experimental MS data according to NIST standard and the theoretical values of $I^{Theor}{}_{SD}$ equation (6) of ions at m/z 75, 73, and 55 of linoleic acid (Figure 7(C).)

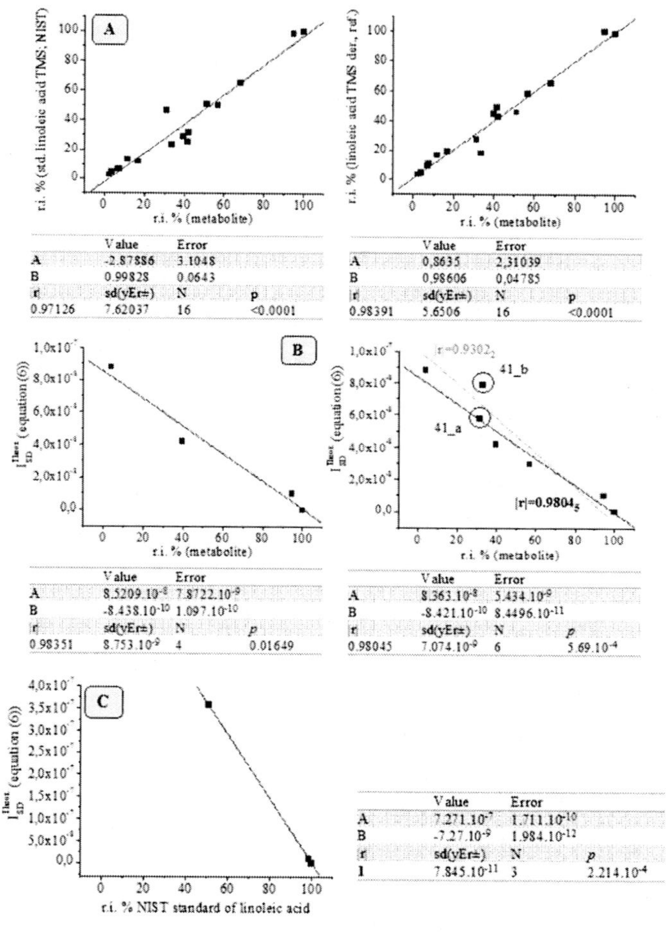

Figure 7. Correlative analysis of mass spectrometric data on metabolite and NIST standard sample of silylated linoleic acid as well as experimental data on the latter analyte according to the study [87] (A); linear relationship between theoretical intensity data ($I^{Theor}{}_{SD}$) of equation (6) and experimental values of metabolite peaks measured (B); and linear relationship between theoretical intensity data ($I^{Theor}{}_{SD}$) of equation (6) and experimental values of peaks measured at m/z 75, 75, and 55 of NIST standard sample of silylated linoleic acid (C) [https://webbook.nist.gov/cgi/inchi?ID=C56259075&Mask=200#Mass-Spec]; chemometrics.

Table 1. ANOVA test data on metabolite and NIST standard spectrum of linoleic acid TMS and GC–MS data on the same analyte according to [87]

Dataset		N	Mean	sd(yEr±)	se(yEr±)	
Data1_metab		16	38.12619	30.59868	7.64967	
Data1_NISTstdLA		16	35.18187	31.42079	7.8552	
	H_0: The means of all selected datasets are equal					
	H_1: The means of one or more selected datasets are different					
Source		DoF	Sum of square	Mean square	F value	*p* value
Model		1	69.3518765	69.3518765	0.07211	0.79013
Error		30	28853.1716	961.772386		
	At the 0.05 level, the population means are not significantly different.					
Dataset		Mean	Difference between means	Simultaneous confidence intervals		Significant at 0.05 level
Data1_metab		38.12619		Lower Limit	Upper Limit	
Data1_NISTstdLA		35.18187	2.94431	-19.4483	25.33693	No
Dataset		N	Mean	sd(yEr±)	se(yEr±)	
Data1_metab		16	36.91485	30.36603	7.36484	
Data1_Ref_[87]		16	35.18187	31.42079	7.8552	
	H_0: The means of all selected datasets are equal					
	H_1: The means of one or more selected datasets are different					
Source		DoF	Sum of square	Mean square	F value	*p* value
Model		1	24.7535838	24.7535838	0.02596	0.87305
Error		30	29562.5201	953.629680		
	At the 0.05 level, the population means are not significantly different.					
Dataset		Mean	Difference between means	Simultaneous confidence intervals		Significant at 0.05 level
Data1_metab		36.91485		Lower Limit	Upper Limit	
Data1_Ref_[87]		35.18187	1.73297	-20.20462	23.67056	No

The results of ANOVA tests confirm the given statement (Table 1). The same statement is true for the data for the Kolmogorov–Smirnov test of the analyte relative intensity parameters. The median test outcome correlates to chemometric parameters discussed, so far (Tables 2 and 3.)

Table 2. Kolmogorov–Smirnov test data on relative intensity data in metabolite and NIST standard variables of silylated linoleic acid of two sample measurable m/z variables of 16 parent and product ions

Dataset	N	Min	Q1	Median	Q3	Max
Data1_metab	16	2.5315	8.70605	36.8004	55.4399	100
Data1_NISTstdLA	16	2.34	7.6025	27.01	50.7125	100
D	Z	Exact probability > $\lvert D\rvert$				
0.1875	0.53033	0.92556				
H_0: F(x) = G(x)						
H_1: F(x) <> G(x)						

At the 0.05 level the two distributions are not significantly different

Table 3. Median test data on relative intensity data in metabolite and NIST standard variables of silylated linoleic acid of two sample measurable m/z variables of 16 parent and product ions

Dataset	N	Min	Q1	Median	Q3	Max
Data1_metab	16	2.5315	8.70605	36.8004	55.4399	100
Data1_NISTstdLA	16	2.34	7.6025	27.01	50.7125	100

Median	χ^2	DF	Probability > χ^2
31.5143	0.5	1	0.4795
H_0: The samples come from the same population			
H_1: The samples come from different populations			
At 0.05 level the two distributions are not significantly different			

3D molecular structural analysis of parent and product ions of LA according to equation (6) study the relationship of $I^{Theor}{}_{SD} = f(\text{r.i.\%})$ yield to the excellent linear relation showing |r| = 0.98351 (Figure 7.) Therefore, there is a correlation between LA metabolomics annotation and stochastic dynamics mass spectrometry, based on equation (6).

Due to the fact that MS ions of silylated LA at m/z 41, 55, and 81 can be attributed to both the organic fragment of the analyte or its silylated residue, the set of possible cations of the product was studied. As shown on Figure 7, the best chemometric |r|-parameter is obtained by distributing MS ions at m/z 41 and 81 to cations 41_a and 81_b, obtaining |r| = 0.9804_5 and six theoretically predicted ions of the product of the analyte under study.

Table 4 summarizes the energetic characteristics of LA ions related to cations at m/z 352, 337, 81, 75, 73, and 41 respectively. The atomic coordinates of the silylated linoleic acid ion and the corresponding theoretical vibrational modes are given in Tables 5 and 6. The data of the molecular dynamics of ions depending on the theoretical level of calculations are shown in Figure 8. Born–Oppenheimer MD data on various theoretical levels together with the theoretical infrared spectra of species in GS are shown in Figures 9 and 10.

Figure 11 shows chemometrics correlation of D_{QC} parameters of equation (6) depending on the theoretical level and ions of silylated LA of the product under study at m/z 41, 73, and 81. The excellent method performance (|r| = 0.9153_1) was obtained showing that two SDD and cc-pVDZ basis sets give qualitatively identical and significant mutually correlated D_{QC} data on silicon-containing fragment ions of LA analyte. The ion at m/z 73 refers to 3D molecular structure of 73_a (see below.) The reliability of the described parameter in the methods M062X/sdd, MP2/SDD, MP2/STO-3G, MP2/cc-pVDZ, and MP2/LANL2DZ can be estimated, including the theoretical values of vibrational frequencies of ions at m/z 41 and 81 at GS and TS (Table 6 and Figure 12). As can be seen, when considering the oscillatory modes GS, sd(yEr±) = 43.76–148.29 are obtained.

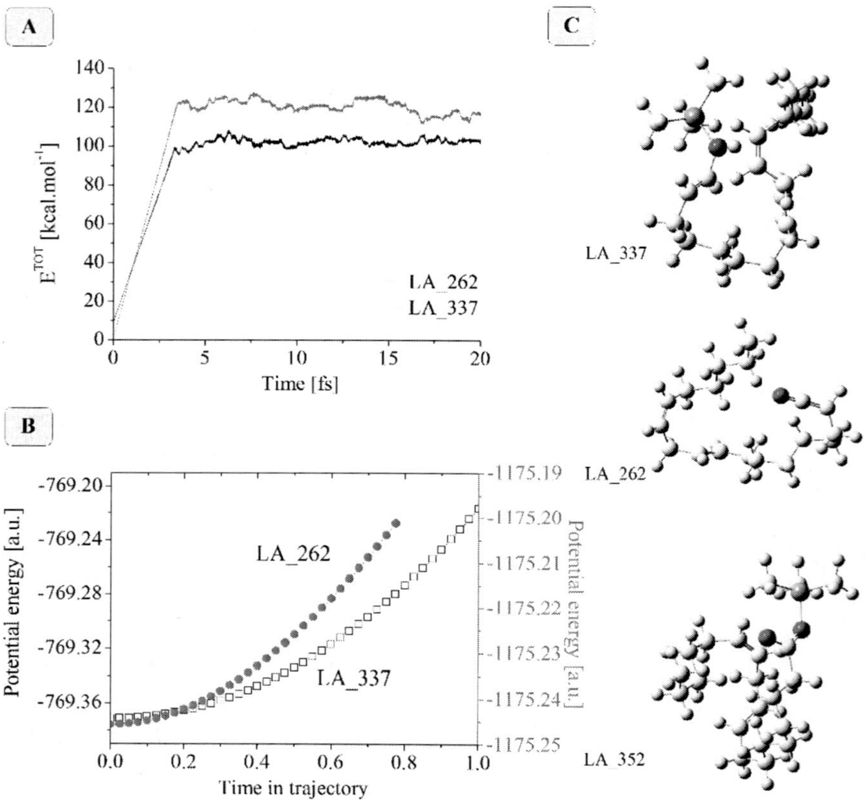

Figure 8. Molecular dynamics data of silylated linoleic acid ions at m/z 352, 337, and 262 as shown in Figure 6; MM2/MD calculations: total energy E^{TOT} [kcal.mol^{-1}] *versus* time [fs] (A); DFT-Born–Oppenheimer MD calculations: potential energy [a.u.] *versus* time in trajectory [fs] (B); and DFT optimized 3D molecular structures of the species (C), respectively.

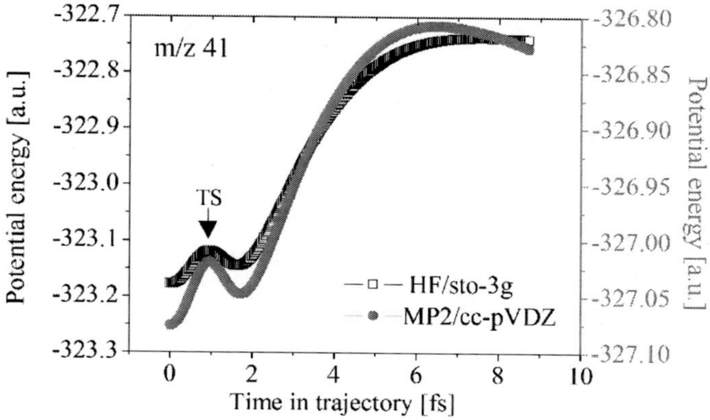

Figure 9. Born–Oppenheimer molecular dynamics data of ion at m/z 41 (form 41_b) depending on the theoretical level; potential energy [a.u.] *versus* time in trajectory [fs]; the transition state point is highlighted.

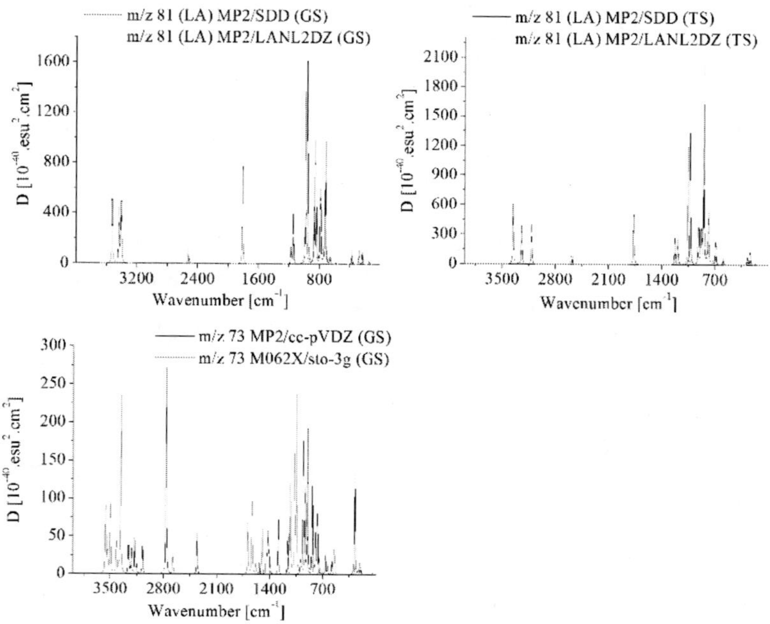

Figure 10. Theoretical vibrational modes in the ground and transition state of ion at m/z 81 (form 81_b) depending on the theoretical level of calculations.

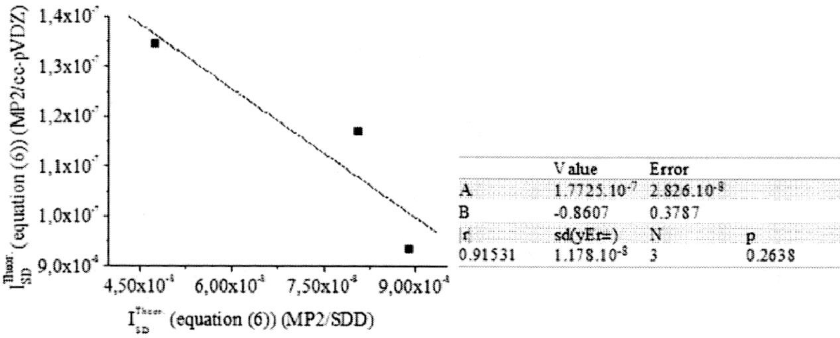

Figure 11. Correlative analysis of equation (6) D_{QC} data of ions at m/z 81 (81_b), 73 (73_b; $[SiH(CH_3)(CH_2CH_3)]^+$)), and 41 (41_b) depending on the theoretical level of calculations; chemometrics.

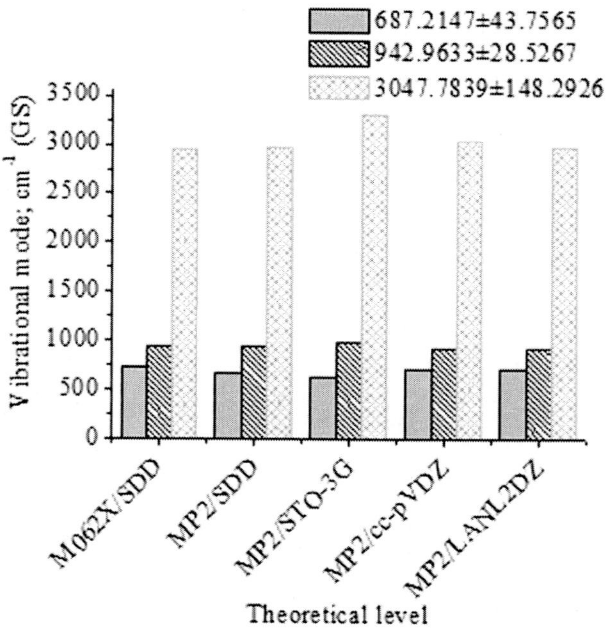

Figure 12. Theoretical vibrational modes in ground state of ion 41_b at different theoretical levels; mean values of the frequencies and standard deviations for five theoretical methods shown.

Table 4. Thermochemistry of observable mass spectrometric ions of silylated linoleic acid as shown in Figure 6 at ground and transition states in gas-phase; equation (6) theoretical intensity (I^{Theor}_{SD}) data

	m/z 352		m/z 337		m/z 81 (81_a)	
	M062X/STO-3G		M062X/STO-3G		M062X/SDD	
	GS	TS	GS	TS	GS	TS
E_{ZPVE}	429.93609	429.48626	407.17056	407.11123	80.44334	78.65740
E_{corr}	0.710920	0.709729	0.677887	0.677019	0.135595	0.132087
H_{corr}	0.711864	0.710673	0.678831	0.677963	0.136540	0.133032
G_{corr}	0.629766	0.629292	0.587500	0.588568	0.096867	0.094897
ε_0	0.685147	0.684430	0.648868	0.648773	0.128195	0.125349
E	-1241.355792	-1241.349972	-1174.567888	-1174.566810	-233.121180	-233.116584
H	-1241.354848	-1241.349028	-1174.566944	-1174.565866	-233.120236	-233.115640
G	-1241.436946	-1241.430409	-1174.658274	-1174.655261	-233.159908	-233.153775
I^{Theor}_{SD}	$8.7778.10^{-8}$		$4.1829.10^{-8}$			
	m/z 81 (81_a)		m/z 55			
	M062X/STO-3G		M062X/STO-3G			
	GS	TS	GS	TS		
E_{ZPVE}	86.61068	85.34348	45.32502	31.84723		
E_{corr}	0.145563	0.143347	0.057305	0.054463		
H_{corr}	0.146507	0.144291	0.058249	0.055407		
G_{corr}	0.106743	0.104660	0.027131	0.024998		
ε_0	0.138023	0.136004	0.053181	0.050752		
E	-230.566430	-230.561501	-188.513368	-188.479326		
H	-230.565486	-230.560557	-188.512424	-188.478382		
G	-230.605250	-230.600187	-188.543542	-188.508791		
I^{Theor}_{SD}	$1.5903.10^{-7}$		$1.22889.10^{-7}$			

E_{ZPVE} – zero-point vibration energy [kcal/mol]; ε_0 – zero-point correction [Hartree/particle]; E_{corr} – thermal correction to energy [Hartree/particle]; H_{corr} – thermal correction to enthalpy [Hartree/particle]; G_{corr} – thermal correction to free energy [Hartree/particle]; E – sum of electronic and thermal energies [Hartree/particle]; H – sum of electronic and thermal enthalpies [Hartree/particle]; G – sum of electronic and thermal free energies [Hartree.$(particle)^{-1}$]

Table 5. Atomic coordinates of mass spectrometric ion of silylated linoleic acid at m/z 352 according to chemical diagrams shown in Figure 6 in ground (GS) and transition (TS) states

m/z 352				m/z 352			
TS				GS			
O	207.392	-0.95522	184.061	O	180.549	-0.77405	170.157
C	185.267	-13.883	0.56517	C	162.401	-126.302	0.43853
O	258.312	-102.034	-0.43145	O	246.291	-106.325	-0.52033
C	0.63943	-233.471	0.38892	C	0.35423	-212.981	0.25072
Si	418.579	-0.01522	-0.39651	Si	406.678	-0.23165	-0.43754
C	-0.38161	-222.489	155.219	C	-0.53586	-217.613	151.938
C	-147.521	-331.235	143.084	C	-164.406	-325.382	139.006
C	-226.754	-326.638	0.10103	C	-251.903	-311.388	0.11951
C	-297.397	-191.036	-0.13435	C	-31.575	-171.285	-0.0242
C	-378.912	-18.892	-145.176	C	-406.158	-161.875	-127.789

Table 5. Continued

m/z 352				m/z 352			
TS				GS			
C	-437.532	-0.48205	-175.381	C	-451.049	-0.16414	-155.946
C	-332.533	0.59927	-20.292	C	-3.376	0.78629	-194.697
C	-208.363	0.40268	-245.812	C	-216.219	0.43864	-235.991
C	-106.713	153.121	-26.485	C	-106.986	144.906	-271.788
C	0.11393	132.032	-170.006	C	0.1558	127.247	-181.938
C	0.45526	207.987	-0.66397	C	0.52762	206.943	-0.82056
C	-0.22385	336.409	-0.19895	C	-0.18662	332.365	-0.33163
C	-0.89504	319.259	119.543	C	-0.67477	316.667	113.676
C	-208.798	22.064	114.749	C	-167.575	199.708	12.885
C	-282.867	206.919	250.414	C	-227.537	187.814	271.187
C	-195.862	145.188	362.192	C	-122.052	159.124	380.329
C	41.117	0.79649	126.882	C	408.316	0.64311	119.815
C	381.956	110.269	-184.661	C	402.022	0.85561	-193.453
C	538.346	-1.321	-0.61627	C	521.669	-168.121	-0.53948
H	106.184	-335.433	0.32734	H	0.7162	-31.374	-0.02091
H	0.19148	-208.873	-0.58677	H	-0.18199	-171.348	-0.61835
H	-0.83575	-122.556	15.352	H	-0.98649	-118.757	167.862
H	0.1418	-234.728	251.231	H	0.09006	-240.938	239.281
H	-216.723	-317.104	227.208	H	-228.242	-317.483	228.053
H	-1.008	-430.083	154.749	H	-117.824	-42.492	140.346
H	-301.534	-407.082	0.12174	H	-331.099	-387.338	0.166
H	-159.339	-347.843	-0.74004	H	-191.779	-333.344	-0.77354
H	-364.218	-169.773	0.71052	H	-375.072	-148.542	0.872
H	-222.977	-110.698	-0.1725	H	-237.123	-0.95167	-0.09726
H	-461.457	-260.919	-137.324	H	-495.044	-224.611	-11.286
H	-315.844	-2.215	-228.861	H	-35.277	-201.467	-215.182
H	-499.069	-0.16309	-0.89996	H	-501.156	0.23981	-0.66709
H	-505.392	-0.54841	-261.871	H	-526.013	-0.16541	-236.654
H	-36.627	162.579	-184.784	H	-362.441	185.174	-187.414
H	-171.451	-0.6052	-266.204	H	-188.002	-0.6139	-245.215
H	-0.6972	153.597	-368.706	H	-0.76249	130.712	-376.681
H	-154.981	249.744	-24.609	H	-147.136	246.397	-262.029
H	0.69612	0.41782	-191.402	H	0.75323	0.38111	-204.358
H	130.706	177.636	-0.04168	H	142.762	181.953	-0.24802
H	0.53119	416.196	-0.13555	H	0.50382	417.896	-0.38464
H	-0.98076	368.166	-0.92568	H	-104.402	354.568	-0.97752
H	-125.437	417.253	153.785	H	-115.418	410.312	145.224
H	-0.14501	284.604	19.171	H	0.19263	300.333	17.905
H	-173.283	122.088	0.81713	H	-117.079	105.857	102.231
H	-280.046	256.075	0.39071	H	-249.541	21.355	0.56941
H	-371.763	144.136	235.612	H	-302.123	107.129	270.773
H	-317.918	305.911	28.251	H	-280.396	280.964	29.552
H	-254.892	131.436	453.432	H	-170.908	142.951	477.016
H	-110.949	209.967	386.171	H	-0.52557	243.057	391.153

m/z 352				m/z 352			
TS				GS			
H	-157.333	0.47412	331.306	H	-0.64216	0.6935	355.824
H	496.355	148.118	130.997	H	501.047	121.836	127.914
H	424.146	0.07163	207.684	H	406.241	-0.05586	203.947
H	321.317	138.895	146.162	H	325.648	135.048	130.383
H	293.689	17.326	-17.118	H	320.994	158.641	-188.026
H	369.334	0.52156	-276.345	H	389.376	0.26128	-284.273
H	468.769	175.371	-197.649	H	496.411	140.139	-201.821
H	538.243	-203.382	0.21133	H	517.554	-229.363	0.36435
H	637.203	-0.85568	-0.65181	H	624.509	-132.948	-0.65884
H	52.404	-187.032	-154.965	H	497.806	-231.129	-139.991
m/z 81(81_b)							
GS				TS			
Si	-0.16331	-0.02919	0.42547	Si	0.14683	0.00641	0.44123
O	-135.574	0.75918	-0.40529	O	135.428	-0.7518	-0.4278
O	0.0602	-156.542	-0.14947	O	-0.04776	151.263	-0.10254
O	137.698	0.69629	-0.24275	O	-144.324	-0.6611	-0.18195
H	0.10482	0.21129	181.644	H	-0.12479	-0.22708	182.175
H	-172.254	151.729	0.14841	H	17.025	-153.288	0.15352
H	-0.6571	-175.664	-0.83269	H	0.637	174.082	-0.8273
H	17.949	0.29787	-106.261	H	-181.488	-0.3137	-105.723
H	211.465	101.838	0.35396	H	-213.505	-103.432	0.35952

Table 6. Theoretical vibrational modes in ground a transition states of silylated linoleic acid product ions at m/z 41 (41_b) and 81 (81_b) shown in Figure 6 at various theoretical levels; vibrational modes [cm^{-1}]

m/z 41 (41_b)	GS			TS		
M062X/sdd	729.1303	948.2756	2954.4265	-1958.6001	1049.9727	5678.0563
MP2/SDD	661.9743	938.9863	2974.5873	-1996.5218	1076.1907	5709.6100
MP2/STO-3G	623.2101	987.9535	3309.2438	-537.9343	562.1308	3471.4526
MP2/cc-pVDZ	715.7728	927.3469	3022.9645	-2031.0745	1058.6087	5740.9716
MP2/LANL2DZ	705.9861	912.2540	2977.6974	-1994.0591	1056.3129	5716.9723
m/z 81 (81_b)	GS			TS		
M062X/STO-3G	41.4604	202.9758	250.2483	-858.1667	150.1476	199.6949
	295.1857	391.8037	420.4058	285.2272	425.3957	577.5011
	454.1197	667.7148	784.3957	658.6981	669.8567	716.2130
	818.5358	849.4131	973.5871	795.9128	867.6924	941.9087
	975.7152	1215.9876	1240.7047	1046.3020	1246.4205	1336.1501
	1877.9382	2749.8126	3889.3967	1803.8233	2798.3652	3654.6582
	3924.5988	3926.9251	4082.9264	3749.0266	3840.6297	4622.1791
MP2/SDD	139.5361	231.8377	278.7000	-575.3889	173.9406	231.2889
	376.8869	586.3674	656.6395	272.5099	597.8156	685.5194
	718.2586	753.0219	780.7778	726.9998	786.0526	806.7905
	855.9692	874.1425	977.5496	845.4240	882.7908	921.1490
	981.2794	1148.6426	1177.1600	1055.1760	1187.8067	1228.6714
	1806.8518	2529.0679	3405.1655	1770.2107	2589.4813	3113.4010
	3412.9506	3434.2851	3523.8882	3244.3521	3353.6936	4104.8387

When studying 3D molecular conformation and electronic structure of MS ion at m/z 73, which is related to $[Si(CH_3)_3]^+$ cation, there are several molecular structural modifications and variations in the localization of the electron charge. Therefore, it is necessary to gain a better understanding of the correlation between the theoretical $I^{Ther}{}_{SD}$ parameter of equation (6) and the experimental data of relative intensity, taking into account the chemical reactions of intramolecular ion rearrangement at m/z 73 depending on the position of the positive charge, taking into account different levels of the theoretical calculations, describing cations 73_a ($[Si(CH_3)_3]^+$) and 73_b ($[SiH(CH_3)(CH_2CH_3)]^+$). Therefore, it is necessary not only to get an idea of the chemical reactivity of the MS species under study at m/z 73 and its 3D molecular and electronic structures, but also to take into account the contribution of errors of the calculations theoretical levels to D_{QC} parameter of equation (6). Due to the fact that the above-mentioned parameter is used to receive data on the theoretical intensity on MS peaks, the results discussed shed light on the reliability of structural analysis as well as predictive power of equation (6) in terms of different levels of accuracy of quantum-chemical calculations.

Figure 13, in context of the previous paragraph, shows DFT optimization of ions 73_a and 73_b depending on the theoretical level. Despite this, the level of the theoretical calculations of ion 73_b is characterized by intraionic charge transfer, due to which the most stable cation $[SiH(CH_3)(CH_2CH_3)]^+$ is formed. Despite the ionic structure, the positive charge is localized on the silicone-center. The difference in the values of the electrostatic potential of the silicon-center is $\Delta ESP = |0.001686|$ a.u. at MP2/cc-pVDZ level of the theory. Due to the previous reason, there is a difference in total energy $\Delta\Delta E^{TOT} = |0.034311|$ a.u. of two species (73_a and 73_b) at the latter level of the theory (Table 7) and $\Delta\Delta E^{TOT} = |0.02982|$ a.u. [37]; thus, both of them can be observed experimentally with a high probability depending on the ionization of MS conditions.

Nevertheless, ion 73_a is thermodynamically preferred to cation 73_b (MP2/cc-pVDZ (this paper) and HF/STO-3G [37] theory level). It is important to note that the calculations of CAS(4,4)/6-31G** show a lack of contribution to the energy of the ions, due to the spin-orbit interaction. Recent results are consistent with data from [67, 68]

However, the values of the deletion energy, which are a strict variational value of the expected value of the wave function of Lewis natural molecular structure, are considered in more detail about the energetics of chemical bonds of species 73a and 73b [54, 55]. Thus, the difference in bond energies in structurally similar compounds is estimated. The energy analysis of the NBO deletion shows the following parameters: E_{del}: –407.526898819, E_{SCF}: –407.526898825 a.u.; energy difference between the given parameters: ΔE = |1.10^{-8}| a.u. (73_a), as well as E*del*: –407.493849149, E_{SCF} = –407.493849153, ΔE = |1.10^{-8}| (73_b) at HF/6-31G* level of theory.

As can be seen, the difference in the energy parameter does not provide a significantly distinguishable criterion between two ionic structures belonging to MS peak at m/z 73.

In connection with these reasons, the natural ionicity parameter (i_{AB}) is calculated taking into account the natural polarization coefficients (Table 8). The following data on Si–C and Si–H bonds in cations related to MS peak at m/z 73 were obtained: $i_{Si\text{-}C}$ = –0.4412 (73_a), –0.3937 (73_b), and $i_{Si\text{-}H}$ = –0.2104 (73_b). At i_{AB}=0, a covalent bond occurs, and at i_{AB} = -1, a pure ionic bond A–B and an ionic hybrid on A-atom occurs. The obtained practically comparable and negative parameters of $i_{Si\text{-}C}$ ions 73_a and 73_b agree with ESP data showing the localization of the charge on the silicon atom. The presence of a Si–H hydride bond in the cationic 73_b, if any, does not affect $i_{Si\text{–}C}$ values. These results suggest that the formation of hydride bonds does not affect the nature of Si–C bonds, despite the fact that ion 73_a has three Si–C bonds, while cation 73_b has only two. In addition, the data obtained suggest that neither their cationic energy parameters nor the natural ionicity parameters are able to quantitatively distinguish two ion structures at m/z 73.

However, what do the described results say about 3D molecular and electronic structures of the species in transition state, due to the fact that D_{QC} parameter of equation (3) and $I^{Theor}{}_{SD}$ values of equation (6) take into account the vibrational modes of MS ions in both the GSs and TSs? To answer this question, it is necessary to describe PESs of the species in the transition state (Figure 14.)

At the same time, both the energy parameters and vibrational TSs modes of cations 73_a and 73_b are investigated at different ab initio and DFT levels of the theory (see also the data in Table 7). As you can see, there are several transient states involved.

It is important to note that the transition state 73_a_TS3 is identical to 73_b_TS2, in addition to the fact that there is a slight difference in the energy data on species 73_a_TS2 and 73_b_TS1 (ΔΔE = |0.056| a.u.) In spite of the similarity of the electronic structural parameters of ions 73_a and 73_b and their energy in GS, there is a close energy difference in the ionic structures of

TS. The given data suggest facilitated intramolecular rearrangement depending on the experimental conditions of MS measurements and environmental factors. They also assume presence of their equilibria,

Obviously, for reliable annotation of metabolites using experimental data on ion intensities at m/z 73, it is necessary to take into account $I^{Ther}{}_{SD}$ parameters of ion 73_a, 73_b and dications of their mixtures (see also data on [37]).

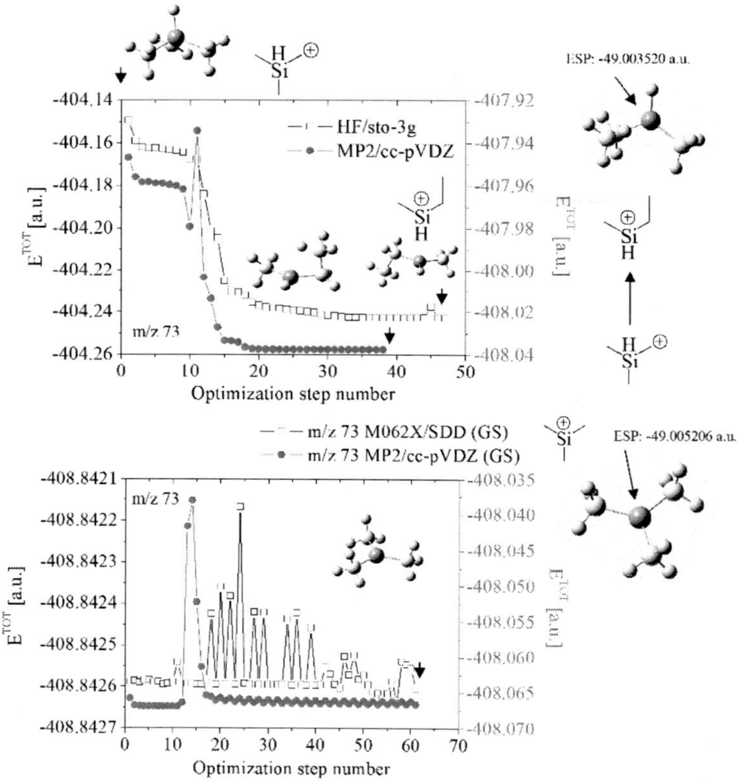

Figure13. *Ab initio* and density functional theory optimization of ion at m/z 73 (forms 73_a ($[Si(CH_3)_3]^+$) and 73_b ($[SiH(CH_3)(CH_2CH_3)]^+$)): total energy, E^{TOT} [a.u.] with respect to optimization step number; chemical diagrams of species; 3D molecular structures of the optimized ions; 2D map of electrostatic potentials at MP2/cc-pVDZ level of theory; and electrostatic potential value of silicon center [a.u.].

Despite, the fact that there is a slight difference in the total energy of ions 73_a and 73_b as well as in their natural iconicity parameters of their D_{QC}, respectively, $I^{Theor}{}_{SD}$ values of equation (6) have markedly different values, however, due to significantly different vibrational properties of the ions. This fact makes it possible to determine 3D geometry of the species studying the experimental and theoretical mass spectrometric variables of the peak at m/z 73 within the framework of the stochastic dynamics theory and model equations (2), (3) and (6) correlatively.

Table 7. Thermochemistry of mass spectrometric ions under study at m/z 41, and 81 of silylated linoleic acid as shown in Figure 6 at ground and transition states in gas-phase, depending on the theoretical level of calculations and ionic structures encompassing form 41_a, 41_b, 81_a, and 81_b; theoretical intensity (I^{Theor}_{SD}) data on equation (6))

	m/z 41		m/z 41		m/z 41	
	41_a		41_a		41_b	
	This work		This work		This work	
	M062X/STO-3G		M062X/sdd		M062X/sdd	
	GS	TS	GS	TS	GS	TS
E_{ZPVE}	43.68388	43.49433	41.20952	39.97164	6.62154	9.61820
E_{corr}	0.074128	0.073130	0.070239	0.067782	0.013531	0.018190
H_{corr}	0.075072	0.074074	0.071183	0.068726	0.014475	0.019135
G_{corr}	0.044095	0.044562	0.040139	0.038685	-0.010092	-0.006263
ε_0	0.069615	0.069313	0.065672	0.063699	0.010552	0.015328
E	-115.432425	-115.432123	-116.763429	-116.699500	-327.601879	-327.546505
H	-115.431480	-115.431179	-116.762485	-116.698556	-327.600935	-327.545561
G	-115.462458	-115.460692	-116.793529	-116.728597	-327.625503	-327.570959
I^{Theor}_{SD}	$5.7414.10^{-8}$					
	m/z 41					
	41_a					
	This work		This work		This work	
	MP2/LANL2DZ		MP2/cc-pVDZ		M062X/STO-3G	
	GS	TS	GS	TS	GS	TS
E_{ZPVE}	6.57022	9.72049	6.67050	9.68290	7.03408	5.76630
E_{corr}	0.013465	0.018352	0.013618	0.018293	0.014228	0.012204
H_{corr}	0.014409	0.019297	0.014562	0.019237	0.015172	0.013148
G_{corr}	-0.010179	-0.006100	-0.010017	-0.006160	-0.009453	-0.012869
ε_0	0.010470	0.015491	0.010630	0.015431	0.011210	0.009189
E	-41.794239	-327.002899	-327.062203	-41.735685	-323.898234	-323.874172
H	-41.793295	-327.001955	-327.061259	-41.734741	-323.897289	-323.873228
G	-41.817883	-327.027351	-327.085838	-41.760138	-323.921915	-323.899245
I^{Theor}_{SD}	$9.1537.10^{-8}$		$9.1538.10^{-8}$		$9.0525.10^{-8}$	
	m/z 41		m/z 81		m/z 81	
	41_b		81_a		81_a	
	This work		This work		This work	
	MP2/sdd		M062X/STO-3G		MP2/SDD	
	GS	TS	GS	TS	GS	TS
E_{ZPVE}	6.54107	9.70079	86.61068	85.34348	80.44334	78.65740
E_{corr}	0.013432	0.018319	0.145563	0.143347	0.135595	0.132087
H_{corr}	0.014376	0.019263	0.146507	0.144291	0.136540	0.133032
G_{corr}	-0.010232	-0.006131	0.106743	0.104660	0.096867	0.094897
ε_0	0.010424	0.015459	0.138023	0.136004	0.128195	0.125349

Table 7. Continued

E	-326.957197	-326.897725	-230.566430	-230.561501	-233.121180	-233.116584
H	-326.956252	-326.896781	-230.565486	-230.560557	-233.120236	-233.115640
G	-326.980861	-326.922175	-230.605250	-230.600187	-233.159908	-233.153775
I^{Theor}_{SD}	$8.9109.10^{-8}$		$1.5903.10^{-7}$			
	m/z 81		m/z 81		m/z 73	
	81_b		81_b		73_a	
	This work		This work		This work	
	MP2/sdd		M062X/STO-3G		MP2/cc-pVDZ	
	GS	TS	GS	TS	GS	TS
E_{ZPVE}	40.95577	40.85418	42.93555	43.43869	68.99024	68.89542
E_{corr}	0.070879	0.070288	0.074904	0.074762	0.117910	0.116989
H_{corr}	0.071823	0.071232	0.075848	0.075707	0.118854	0.117933
G_{corr}	0.036766	0.036924	0.038480	0.040751	0.076525	0.076954
ε_0	0.065267	0.065105	0.068422	0.069224	0.109943	0.109792
E	-516.527531	-516.524203	-511.436474	-511.430890	-407.949409	-407.949932
H	-516.526587	-516.523258	-511.435530	-511.429946	-407.948465	-407.948987
G	-516.561644	-516.557566	-511.472898	-511.464902	-407.990794	-407.989967
I^{Theor}_{SD}	$8.0656.10^{-8}$		$2.959.10^{-8}$			
	m/z 73		m/z 75		m/z 75	
	73_b					
	This work		This work		This work	
	MP2/cc-pVDZ		HF/STO-3G		MP2/cc-pVDZ	
	GS	TS (73_b_TS1)	GS	TS	GS	TS
E_{ZPVE}	69.68146	69.30394	63.56044	63.56044	55.47857	55.43313
E_{corr}	0.117745	0.116878	0.106136	0.106136	0.095355	0.094421
H_{corr}	0.118690	0.117822	0.107080	0.107080	0.096299	0.095365
G_{corr}	0.081383	0.080325	0.073359	0.073359	0.057206	0.058814
ε_0	0.111044	0.110443	0.101290	0.101290	0.088411	0.088338
E	-407.919758	-407.919103	-437.933724	-437.933724	-443.881683	-443.882609
H	-407.918814	-407.918159	-437.932780	-437.932780	-443.880739	-443.881664
G	-407.956121	-407.955656	-437.966501	-437.966501	-443.919832	-443.918216
	m/z 73		m/z 73		m/z 73	
	73_a		73_b		73_a	
	[37]		This work		This work	
	M062X/STO-3G		MO62X/STO-3G		MO62X/STO-3G	MO62X/SDD
	GS	TS (73_a_TS1)	GS	TS	TS (73_a_TS2)	TS (73_a_TS2)
E_{ZPVE}	78.53423	77.82701	78.04071	76.98356	78.49353	71.00597
E_{corr}	0.131361	0.130787	0.131025	0.128281	0.130999	0.119332
H_{corr}	0.132305	0.131731	0.131969	0.129225	0.131943	0.120276
G_{corr}	0.096253	0.091193	0.093807	0.094142	0.096246	0.084070
ε_0	0.125152	0.124025	0.124366	0.122681	0.125087	0.113155
E	-404.143528	-404.144788	-404.111591	-404.086694	-404.142491	-408.720255
H	-404.142584	-404.143843	-404.110647	-404.085750	-404.141547	-408.719311
G	-404.178636	-404.184382	-404.148809	-404.120833	-404.177244	-408.755517

The following MS results and method performances were used to extend the chemometric approach to the identification and annotation of carbohydrates. Figures 15–17 also depict experimental GC–MS spectra of metabolites and the corresponding spectra of NIST standards and a set of reference measurements [87].

Table 8. Natural bond orbital analysis of silicon containing ions 73_a and 73_b at HF/6-31G* level of theory

Occupancy	Bond orbital	Natural polarization coefficients (c_i)	Hybrids
73_a			
1.97450 (27.94%)	BD(1)Si1–C2	0.5286*Si	$1s(33.15\%)p^{1.96}(65.03\%)d^{0.05}(1.82\%)$
(72.06%)		0.8489* C	$2s(24.99\%)p^{3.00}(74.91\%)d^{0.00}(0.10\%)$
1.97472 (27.77%)	BD(1)Si1–C3	0.5270*Si	$1s(33.27\%)p^{1.95}(64.89\%)d^{0.06}(1.83\%)$
(72.23%)		0.8499* C	$3s(25.27\%)p^{2.95}(74.63\%)d^{0.00}(0.10\%)$
1.97473 (27.79%)	BD(1)Si1–C4	0.5272*Si	$1s(33.29\%)p^{1.95}(64.87\%)d^{0.06}(1.83\%)$
(72.21%)		0.8498*C	$4s(25.17\%)p^{2.97}(74.73\%)d^{0.00}(0.10\%)$
73_b			
1.97126 (30.31%)	BD(1)Si1–C2	0.5506*Si	$1s(35.26\%)p^{1.79}(63.21\%)d^{0.04}(1.53\%)$
(69.69%)		0.8348* C	$2s(21.80\%)p^{3.58}(78.09\%)d^{0.01}(0.12\%)$
1.97910 (29.59%)	BD(1)Si1–C3	0.5440*Si	$1s(34.15\%)p^{1.88}(64.24\%)d^{0.05}(1.60\%)$
(70.41%)		0.8391* C	$3s(23.91\%)p^{3.18}(75.99\%)d^{0.00}(0.11\%)$
1.97019 (39.48%)	BD(1)Si1–H5	0.6283*Si	$1s(30.28\%)p^{2.24}(67.94\%)d^{0.06}(1.78\%)$
(60.52%)		0.7779* H	5s(100.00%)

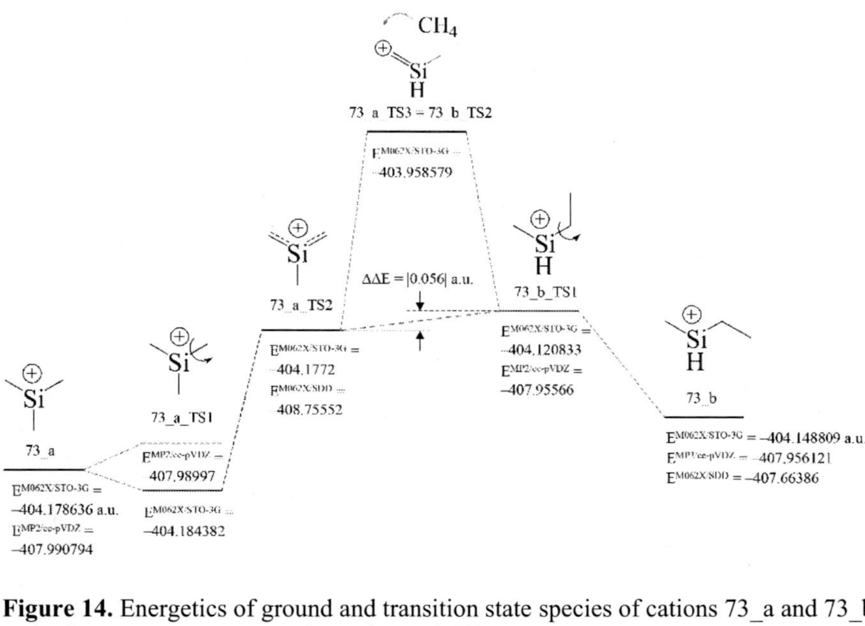

Figure 14. Energetics of ground and transition state species of cations 73_a and 73_b assigning to mass spectrometric peak at m/z 73.

An analyte with RT = 16.04 min was assigned to 1,5-anhydrohexitol correlated with the data at the reference [87] with statistical significance |r| = 0.9157 (Figure 18). The correlation with measurable variables of the standard NIST sample of DB1 yields |r| = 0.9259, while analysis of measurands of the latest standard measurements and reference data [87] shows excellent and accurate values (|r| = 0.99987.) The lower |r|-parameter of the metabolomic annotation of our metabolite is to say that its annotation to 15ADG is not without controversy; furthermore, the standard sample of 15ADG does not have the characteristic MS peak at m/z 659. To solve the described problem a correlative analysis of product ions of metabolite and NIST standard sample of (3R,4S,5R,6R)-3,4,5-tris((trimethylsilyl)oxy)-6-(((trimethylsilyl)oxy) methyl)tetrahydro-2H-pyran-2-one [https://webbook.nist.gov/cgi/inchi?ID= C55515292&Mask=200] was carried out, studying peaks at m/z 45, 73, 91, 103 and 129; thus, causing for increasing in performances to |r| = 0.98278.

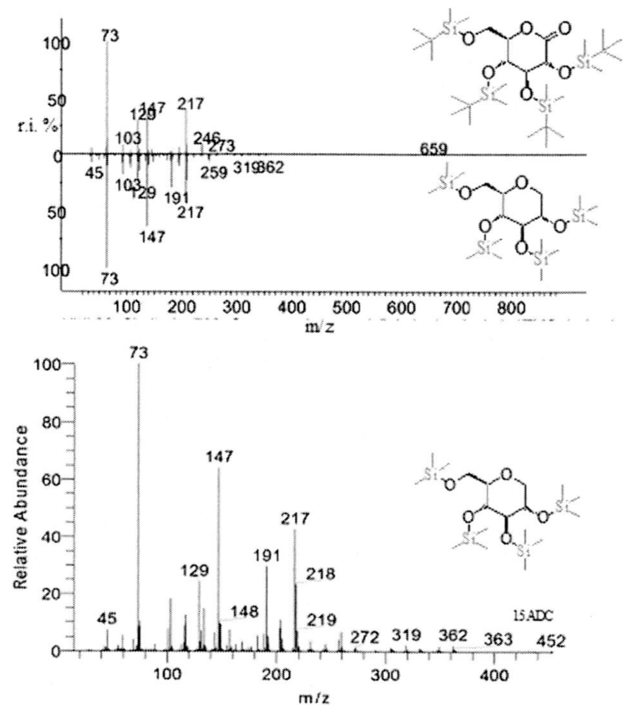

Figure 15. Experimental GC–MS spectra of metabolite at RT = 16.04 min, NIST standard sample of silylated 1,5-anhydro-D-glucitol (DB1) and data on the analyte according to reference [87]; chemical diagrams of *silylated* carbohydrates.

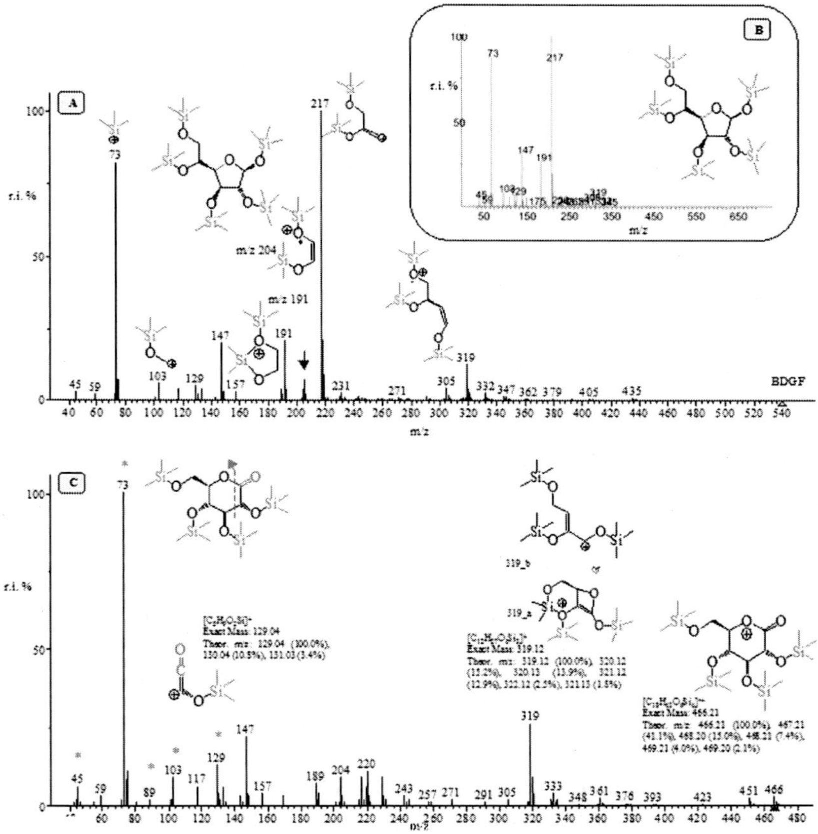

Figure 16. Experimental GC–MS spectra of NIST standard sample of silylated β-D-galactofuranose (A), metabolite (B); and the standard NIS spectrum of silylated gluconolactone (3R,4S,5R,6R)-derivative (C); theoretical m/z data and intensity ratios of isotope shape of peaks of analyte ions; chemical diagrams of *silylated* carbohydrate and its product ions.

Due to these reasons, the latter metabolite is attributed to tetra-O-*t*-butyldimethylsilyl-D-glucono-1,5-lactone. The peak at m/z 659 is attributed to its sodium adduct in accordance with results of the papers [88–90].

The analysis of β-D-galactofuranose results to |r| = 0.9925 taking into account measurable variables of metabolite and NIST spectrum of the standard sample (Figure 19.) The correlation between measurands and the results obtained in the reference [87] in which the metabolite is attributed to assigning metabolite to 1-methyl-α-D-glucofuranose shows $|r| = 0.9998_9$.

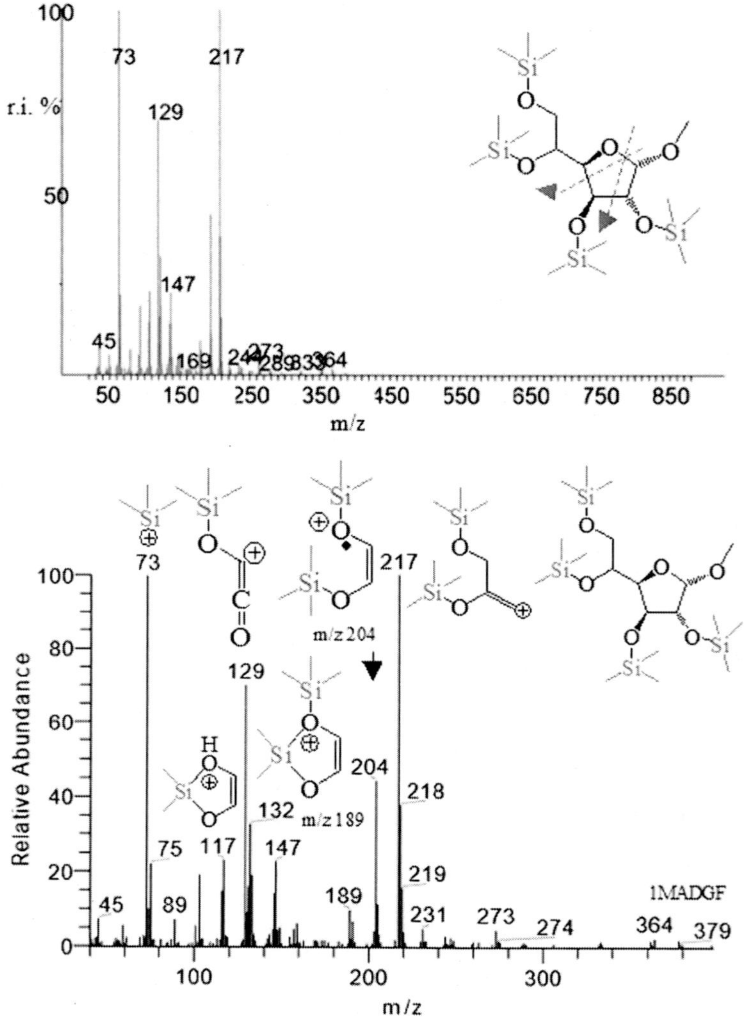

Figure 17. Experimental GC–MS spectra of metabolite (A) and data on silylated 1-methyl-α-D-glucofuranose according to work [87]; chemical diagrams of *silylated* carbohydrate and its product ions.

Figure 20 presents experimental MS data on silylated α-D-arabinofuranose and α-D-glucopyranoside. The correlation analysis of the obtained data on the intensity of peaks at m/z 339, 217, 147, and 73 in the former case, as well as at m/z 217, 204, and 133 in α-D-glucopyranoside in comparison with results from the papers [91, 92] resulted to |r| = 0.988882 and 0.996 (Figure 21.)

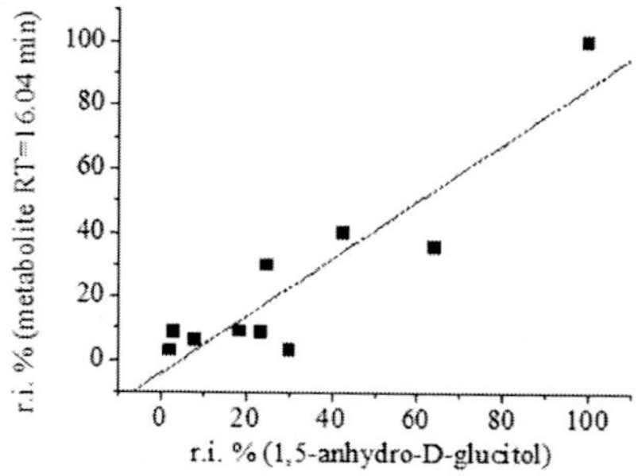

	Value	Error		
A	-4.17328	5.99933		
B	0.89991	0.13963		
|r|	sd(yEr±)	N	p	
0.9157	12.78934	10	1.99377.10-4	

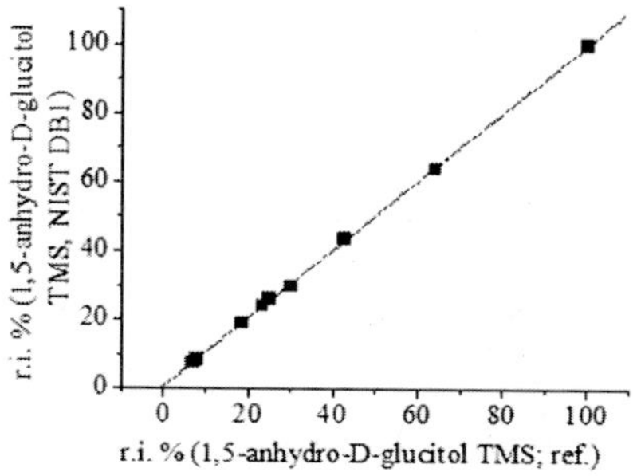

	Value	Error	
A	0.62731	0.27008	
B	0.99292	0.00596	
|r|	sd(yEr±)	N	p
0.99987	0.5044	9	<0.0001

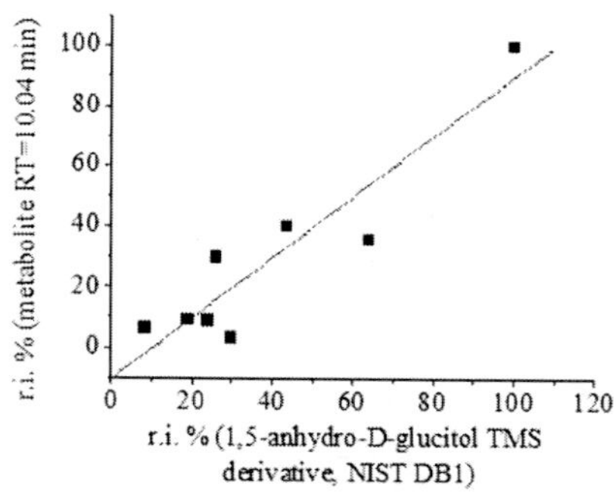

	Value	Error	
A	-10.41478	8.03985	
B	1.00159	0.16683	
|r|	sd(yEr±)	N	p
0.9259	13.10604	8	9.612.10-4

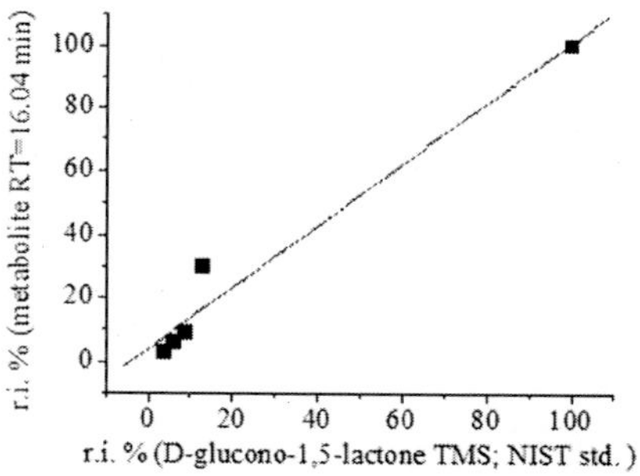

	Value	Error	
A	3.98567	4.77801	
B	0.9698	0.10526	
|r|	sd(yEr±)	N	p
0.98278	8.6902	5	0.0027

Figure 18. Correlative analysis between mass spectrometric measurable variables of metabolite at RT = 16.04 min and NIST standard samples of *silylated* 1,5-anhydro-D-glucitol according to reference [87] and the standard NIST sample of DB1 together with correlation between metabolite data at RT = 16.04 and NIST standard of (3R,4S,5R,6R)-3,4,5-tris((trimethylsilyl)oxy)-6-(((trimethylsilyl)oxy)methyl) tetrahydro-2H-pyran-2-one [https://webbook.nist.gov/cgi/inchi?ID=C55515292&Mask=200]; chemometrics.

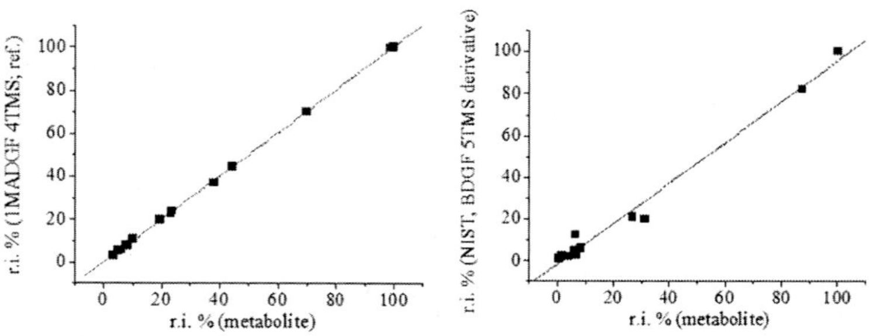

	Value	Error		
A	0.19149	0.20167		
B	1.00059	0.00437		
|r|	sd(vEr±)	N	p	
0.99989	0.53431	14	<0.0001	

	Value	Error		
A	-1.52478	1.39981		
B	0.97066	0.03604		
|r|	sd(vEr±)	N	p	
0.9925	4.13263	13	<0.0001	

Figure 19. Correlative analysis between mass spectrometric measurable variables of metabolites and NIST standard samples of silylated β-D-galactofuranose [https://webbook.nist.gov/cgi/inchi?ID=C7045525&Mask=200#Mass-Spec] and results of 1-methyl-α-D-glucofuranose, 2,3,5,6-tetratakis-*O*-(trimethylsilyl)-derivative according to reference [87]; chemometrics.

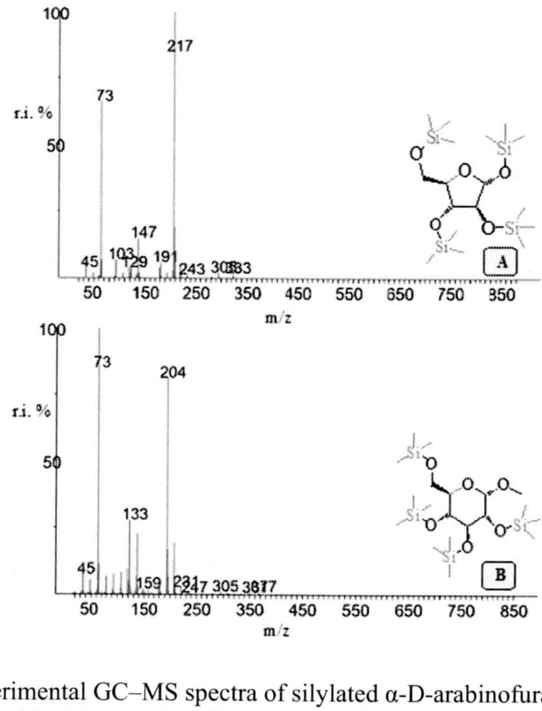

Figure 20. Experimental GC–MS spectra of silylated α-D-arabinofuranose (A) and α-D-glucopyranoside (B); chemical diagrams of analytes.

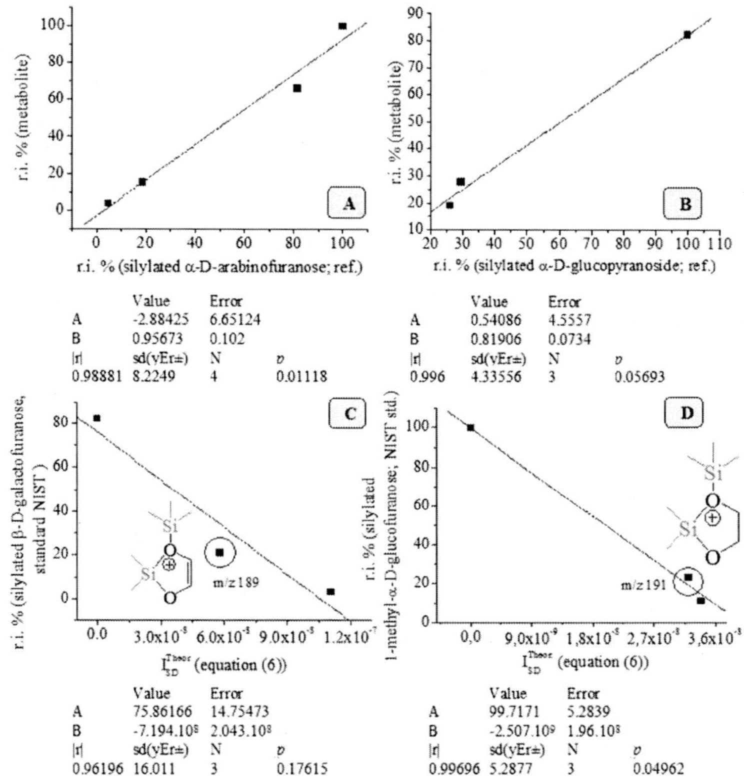

Figure 21. Correlative analysis between an experimental set of GC–MS measurements of metabolites and the results from the reference samples of α-D-arabinofuranose (A); α-D-glucopyranoside (B) according to [91, 92]; β-D-galactofuranose (C); and 1-methyl-α-D-glucofuranose (D); chemical diagrams of ion at m/z 189 and 191; chemometrics.

The comparative analysis of the characteristic MS ions of silylated carbohydrates β-D-galactofuranose and 1-methyl-α-D-glucofuranose (Figures 16 and 17) within the range of m/z 40–250 shows a number of common characteristic MS peaks at 45, 73, 75, 103, 117, 129, 147, 204 and 217; however, excluding from the peaks at m/z 191 and 189. As can be seen, MS peak at m/z 191 is characteristic of silylated β-D-galactofuranose, while the ion at m/z 189 is observed in the mass spectrum of α-D-glucopyranoside. In this regard, the theoretical analysis according to equation (6) involves the above-mentioned ion together with MS peak at m/z 117, due to the fact that within the range of m/z measured values the largest number of MC peaks of both analytes is observed. The energy of the discussed species together with $I^{Theor}{}_{SD}$ parameters of equation (6) are given in Table 9. Figure 21 shows the correlation between the theoretical and experimental mass spectrometric intensity data of silylated β-D-galactofuranose and 1-methyl-α-D-glucofuranose. As can be seen, the excellent chemometric parameters $|r|$ = 0.96196 and 0.99696 of β-D-galactofuranose and 1-methyl-α-D-glucofuranose were received that correlate between the theoretical and experimental MS data of ions at m/z 191, 73, and 45 in the previous case, and m/z 189, 147, and 73, in the study of 1-methyl-α-D-glucofuranose. The peak data at m/z 45, 73, and 147 are common for silylated metabolites as reported (2023) [37].

Table 9. Thermochemistry of the mass spectrometric ions of silylated 1-methyl-α-D-glucofuranose and β-D-galactofuranose (m/z = 191) as shown in Figures 16 and 17 at ground and transition states in gas-phase; theoretical intensity ($I^{Theor}{}_{SD}$) data on equation (6)

	1MADGF		BDGF		1MADGF	
	m/z 117		m/z 189		m/z 189	
	GS	TS	GS	TS	GS	TS
E_{ZPVE}	89.08537	89.19255	180.16313	179.97231	163.09509	162.97248
E_{corr}	0.150588	0.150086	0.303225	0.302218	0.275632	0.274600
H_{corr}	0.151532	0.151030	0.304169	0.303162	0.276576	0.275545
G_{corr}	0.108427	0.108916	0.243443	0.243976	0.216835	0.218238
ε_0	0.141967	0.142137	0.287108	0.286804	0.259909	0.259713
E	-590.186408	-590.174494	-995.280980	-995.267545	-994.074436	-994.066229
H	-590.185464	-590.173550	-995.280036	-995.266601	-994.073492	-994.065285
G	-590.228569	-590.215664	-995.340763	-995.325786	-994.133234	-994.122592
$I^{Theor}{}_{SD}$	$5.88775.10^{-8}$		$5.8066.10^{-8}$		$3.39638.10^{-8}$	

3. Experimental

The experiment on tomatoes was conducted in Moscow region (Russia) between 2022 and 2023. The coordinates are 55°7′27′′′ N latitude and 37°56′55′′′ E longitude, moderate continental climate. To evaluate the effect of the growing conditions on leaf metabolism, tomato plants of tomato cultivar 'Michnevsky' were grown in pots with soil containing earthworms and in pots with the same soil but without earthworms. The soil in the pots was sod-podzolic medium loamy slightly humusified. The pots were located on an open experimental site, watering was carried out as needed. Twelve replicates were used for each soil group.

3.1. Plant Preparation and Extraction

The leaf samples with a petiole were taken in morning, in phase of mass flowering, from middle part of the plant, one leaf from the middle part of the plant (Figure 22). The leaves from ten plants were examined.

Figure 22. The place where the leaf samples were taken and the type of leaf with a petiole are shown.

The freshly harvested leaves were crushed in an IKA (Germany) pulse action MultiDrive basic crusher with glass balls. For the planned analyses, 0.5 g of the crushed fraction of tomato leaves was extracted with 10 ml of pure methanol for 24 hours. The extracts were centrifuged at 4000 g on Rotofix 32 A (Hettich, Germany) for 20 min. The supernatant was used for measurements.

3.2. Metabolic Analysis by Gas Chromatography–Mass Spectrometry

The metabolites analysis was fulfilled using the method of gas chromatography–mass spectrometry *via* GC–MS chromatograph JMS-Q1050GC (JEOL Ltd., Japan). Capillary column DB-5HT (Agilent, USA) (length 30 m, inner diameter – 0.25 mm, the film thickness – 0.52 μ m, and gas-carrier – helium) was used. The temperature gradients during the analysis were within the range of 40°C to 280°C, the injector and interface temperature was 250°C, while the ionic source to 2.0 $mL.min^{-1}$, split-flow injection mode, with the sample injected in volume 2 mcl of the evaporated extract. The analysis was held for 60 min. The derivation was held using silylation reagent (BSTFA) following the method, detailed in [88, 93–95]. The identification of the substances was done according to NIST-15 National Institute of Standards and Technology (USA) retention behavior and mass spectra (DB1). The scanning range was 33–900 m/z, while the identification of substances credibility was within 75–98%. The study uses the following databases of GS–MS spectra of metabolites: Standard NIST mass spectrum of silylated 9,12 - octadecadienoic acid (Z,Z) [https://webbook.nist.gov/cgi/inchi?ID=C56259075&Mask=200#Mass-Spec]; silylated β-D-galactofuranose [https://webbook.nist.gov/cgi/inchi?ID=C7045525&Mask=200#Mass-Spec]; and silylated gluconolactone (3R,4S,5R,6R)-derivative [https://webbook.nist.gov/cgi/inchi?ID=C55515292&Mask=200].

3.3. Chemometrics

Software **R4Cal Open Office STATISTICs for Windows 7 was used [https://extensions.openoffice.org/en/project/r4calc-r-statistics-ooocalc].** Statistical significance was examined *via* *t*-test. The model fit was obtained by F-test. **ANOVA was utilized. Nonlinear fitting of experimental data was carried out *via* the searching method based on** Levenberg-Marquardt algorithm [96–101]. Nonparameteric two sample Kolmogorov–Smirnov [102], Wilcoxon–Mann–Whitney [103], and Mood's median [104] tests were also applied. ProteoWizard 3.0.11565.0 (2017) [https://proteowizard.sourceforge.io/download.html], mMass 5.0.0 [http://www.mmass.org/download/old.php], QuanBrowser 2.0.7 (Thermo Fischer Scientific Inc.), AMDIS 2.71 (2012) software [https://chemdata.nist.gov/mass-spc/amdis/downloads/AMDIS_Installer-17.zip], and NIST Search Software 2.0 [https://chemdata.nist.gov/dokuwiki/lib/exe/fetch.php?media=chemdata:nist17:nist17demo.zip] were used.

Conclusion

Throughout this chapter, our emphasis has been on gas chromatographic analysis coupled with mass spectrometry as a powerful approach to identify and annotate reliably metabolites of methanol extracts of tomato leaves grown with or without presence of soil earthworms. Forty-nine metabolites were determined, 29 substances of which belong to carbohydrates and their derivatives; as well as 20 substances that are organic, fatty, and amino acids. A major difference was found in the accumulation of eight carbohydrates: maltitol, 2-α-mannobiose, myo-inositol, 1.5-anhydroglucitol, arabinofuranose, methyl-a-D-glucofuranoside, methyl-galactoside, and allofuranose, as well as linoleic acid. GC–MS analysis proved that earthworms affect the metabolic profile of the leaves. The increased linoleic acid content indicates the activation of plant protective processes. In addition, it can be stated that tomato plants grown in the soil with a lack of earthworms possess greater resistance [105]. The scarce data on the described issue, reported so far, however, assumes further systematic examining of the influence of soil earthworms on tomato leaves metabolism.

Since the aims of the study are twofold, the concluding section is divided further by the research topic of the chapter into two parts, the first one of which is already highlighted. In the second part of the conclusion, we will address aspects of the significant contribution of stochastic dynamics mass spectrometry to the molecular structural determination of tomato leaf metabolites based on the following results:

A. Excellent to exact method performances $|r| = 1–0.9835_1$ were achieved by comparing the functional relationship $I^{Theor}_{SD} = f(r.i.\%)$ between theoretical intensity data on equation (6) and experimental relative abundance measurands of analyte parent and product mass spectrometric peaks of linoleic acid; and
B. The enantiomer of 1-methyl-α-D-glucofuranose *via* the relationship $I^{Theor}_{SD} = f(r.i.\%)$ and equation (6) at a level of reliability of $|r| = 0.9969_6$ were determined.

The results, highlighted above have argued that the stochastic dynamics mass spectrometric method provides highly reliable-to-exact determining of metabolite molecular structures even examining structurally very similar analytes, particularly, focusing the reader's attention on carbohydrate enantiomers, which are characterized by virtually similar mass spectrometric patterns, thus producing frequently comparable method performances, examining mass spectra of plant extracts and measurands of standard analytes. The decrease in probability of reliable identification of metabolites via only omics-methods that follow from both various specific conformational changes of molecular structures of analytes and their isomers, and different experimental conditions of measurements of the standard mass spectra of the reference samples, is described in the present chapter based on the data of analysis of β-D-galactofuranose yielding to $|r| = 0.9925$ correlating measurable variables of the metabolite and standard sample of NIST 15 database as well as the results from metabolite and reference [87] presenting the same metabolite to 1-methyl-α-D-glucofuranose at excellent to exact performances, showing $|r| = 0.9998_9$. Even though, perhaps, the above-mentioned statement could be found as objectionable, it should be further sketched that targets of metabolomics could often be chemically indistinct from reagents used to the metabolomics extract. Further, there could be incomplete chemical derivatization of metabolites; multiple analytes *per* mass spectrometric peak, due to degradation or competitive side reactions, as well as; thus, causing for at about 5–15% effectively assigned mass spectrometric massive of metabolites *via* omics-methods [106, 107]. Therefore, it is especially noticeable in theoretical and experimental designs of omics-based identification and annotation of plant metabolites, that, frequently only omics-methods are partially applicable to solve the problem of identification and molecular structural analysis of the extracted analytes. A significant feature in the presented context is built into complementary employment in different theoretical methods in order to provide reliable identification and structural analysis of metabolite molecules; thus, achieving the high standards of the analytical mass spectrometric based determination using quantitative criteria of reliability of chemometrics and statistical methods. The excellent-to-exact method performances reported, herein, from both the omics- and stochastic dynamics mass spectrometric method lend support to the view that their complementary employment in identification and annotation of plant metabolites represents an important and prospective methodological alternative to analyze reliably molecular structures of metabolites to purely automated database metabolomics approaches, applied routinely to the analytical practice.

Acknowledgments

The authors would like to thank the participants of the Strogoorganic online gardening school for growing experimental tomato plants and their caring; and Faculty of Science at Gharyan University for experimental analysis (April 2017).

Conflict of Interest

The authors declare no conflict of interest.

Data Availability

The data presented in this study are available on request from the corresponding author (S.M.)

References

[1] FAOSTAT: Crop and livestock products Food Agric. Organ. United Nations (2023) [https://www.fao.org/faostat/en/#data].

[2] Hirai, M., Yano, M., Goodenowe, D., Kanaya, S., Kimura, T., Awazuhara, M., Arita, A., Fujiwara, T., Saito, K. Integration of transcriptomics and metabolomics for understanding of global responses to nutritional stresses in *Arabidopsis thaliana*. *Proc. Nat. Acad. Sci. USA* 2004, *101*, 10205–10210.

[3] Bölling, C., Fiehn, O. Metabolite profiling of *Chlamydomonas reinhardtii* undernutrient deprivation. *Plant Physiol*. 2005, *139*, 1995–2005.

[4] Nikiforova, V., Gakiere, B., Kempa, S., Adamik, M., Willmitzer, L., Hesse, H., Hoefgen, R. Towards dissecting nutrient metabolism in plants: a systems biology case study on sulphur metabolism. *J. Exp. Bot.* 2004, *55*, 1861–1870.

[5] Urbanczyk-Wochniak, E., Fernie, A. Metabolic profiling reveals altered nitrogen nutrient regimes have diverse effects on the metabolism of hydroponically-grown tomato (*Solanum lycopersicum*) plants. *J. Exp. Bot.* 2005, *56*, 309–321.

[6] Hernandez, G., Ramirez, M., Valdes-Lopez, O., Tesfaye, M., Graham, M., Czechowski, T., Schlereth, A., Wandrey, M., Erban, A., Cheung, F., Wu, H., Lara, M., Town, C., Kopka, J., Udvardi, M., Vance, C. Phosphorus stress in common bean: root transcript and metabolic responses. *Plant Physiol*. 2007, *144*, 752–767.

[7] Kusano, M., Fukushima, A., Redestig, H., Saito, K. Metabikimic approaches toward understanding nitrogen metabolism in plants. *J. Exp. Bot*. 2011, 62, 1439–1453.

[8] Takahashi, H., Imamura, T., Miyagi, A., Uchimiya, H. Comparative metabolomics of developmental alterations caused by mineral deficiency during in vitro culture of *Gentiana triflora*. *Metabolomics* 2012, *8*, 154–163.

[9] Amtmann, A., Armengaud, P. Effects of N, P, K and S on metabolism: New knowledge gained from multi-level analysis. *Curr. Opin. Plant Biol.* 2009, *12*, 275–283.

[10] Varki, A. Biological roles of glycans. *Glycobiol.* 2017, *27*, 3–49.

[11] Feussner, I. The lipoxygenase-dependent degradation of storage lipids; In Trends in Plant Science; Feussner, I., Kuhn, H., Wasternack, C. (eds.) Vol. 6; 2001, pp. 262–267.

[12] Dobson, G. Gas chromatography-mass spectrometry of conjugated linoleic acids and metabolites. Advances in conjugated linoleic acid research. AOCS Publishing; 2020, pp. 13–36.

[13] Wei, P., Hao, L., Thomas, S., Buchberger, A., Steinke, L., Marker, P., Ricke, W., Li, L. Urinary amine metabolomics characterization with custom 12-plex Isobaric DiLeu labeling. *J. Am. Soc. Mass Spectrom.* 2020, *31*, 1854–1860.

[14] Zhao, Y., Zhang, Y., Yang, H., Xu, Z., Li, Z., Zhang, Z., Zhang, W., Deng, J. A comparative metabolomics analysis of phytochemcials and antioxidant activity between broccoli floret and by-products (leaves and stalks). *Food Chem.* 2024, *443*, 138517.

[15] Ivanova, B. Special Issue with Research Topics on Recent Analysis and Applications of Mass Spectra on Biochemistry. *Int. J. Mol. Sci.* 2024, *25*, 1995.

[16] Blaženović, I., Kind, T., Sa, M., Ji, J., Vaniya, A., Wancewicz, B., Roberts, B., Torbašinović, H., Lee, T., Mehta, S., Showalter, M., Song, H., Kwok, J., Jahn, D., Kim, K., Fiehn, O. Structure annotation of all mass Sspectra in untargeted metabolomics. *Anal. Chem.* 2019, *91*, 2155–2162.

[17] Wang J. et al. Mass spectrometry for structural elucidation and sequencing of carbohydrates. TrAC Trends Anal. Chem. 2021, *144*, C. 116436.

[18] Domon, B., Costello, C. E. A Systematic nomenclature for carbohydrate fragmentations in FAB-MS/MS spectra of glycoconju-gates. *Glycoconjugate J.* 1988, *5*, 397–409.

[19] Stephens, E., Maslen, S., Green, L., Williams, D. Fragmentation characteristics of neutral N-linked glycans using a MALDI-TOF/TOF tandem mass spectrometer. *Anal. Chem.* 2004, *76*, 2343–2354.

[20] Spina, E., Romeo, D., Impallomeni, E., Garozzo, D., Waidelich, D., Glueckmann, M. New fragmentation mechanisms in matrix-assisted laser desorption/ionization time-of-flight/time-of-flight tandem mass spectrometry of carbohydrates. *Rapid Comm. Mass Spectrom.* 2004, *18*, 392–398.

[21] Kailemia, M., Ruhaak, L., Lebrilla, C., Amster, I. Oligosaccharide analysis by mass spectrometry: A review of recent developments. *Anal Chem.* 2014, *86*, 196–212.

[22] Zhang, H., Gou, X., Ma, L., Zhang, X., Qu, J., Wang, X., Huang, W., Yan, S., Zhang, X., Xue, J., Xum, S. Reveal the kernel dehydration mechanisms in maize based on proteomic and metabolomic analysis. *BMC Plant Biol.* 2024, *24*, 15.

[23] Wang, R., Du, C., Gu, G., Zhang, B., Lin, X., Chen, C., Li, T., Chen, R., Xie, X. Genome-wide identification and expression analysis of the ADH gene family under diverse stresses in tobacco (*Nicotiana tabacum L.*) *BMC Genomics* 2024, *25*, 13.

[24] Paglia, G., Smith, A., Astarita, G. Ion mobility mass spectrometry in the omics era: Challenges and opportunities for metabolomics and lipidomics. *Mass Spec Rev.* 2022, *41*, 722–765.

[25] Chang, H., Colby, S., Du, X., Gomez, J., Helf, M., Kechris, K., Kirkpatrick, C., Li, S., Patti, G., Renslow, R., Subramaniam, S., Verma, M., Xia, J., Young, J. A practical guide to metabolomics software development. *Anal. Chem.* 2021, *93*, 1912–1923.

[26] Johnson, C., Ivanisevic, J., Benton, H., Siuzdak, G. Bioinformatics: The next frontier of metabolomics. *Anal. Chem.* 2015, *87*, 147−156.

[27] De Livera, A., Dias, D., De Souza, D., Rupasinghe, T., Pyke, J., Tull, D., Roessner, U., McConville, M., Speed. T. Normalizing and integrating metabolomics data. *Anal. Chem.* 2012, *84*, 10768−10776.

[28] Mak, T., Goudarzi, M., Laiakis, E., Stein, S. Disparate metabolomics data reassembler: A novel algorithm for agglomerating incongruent LC–MS metabolomics datasets. *Anal.Chem.*2020, *92*, 5231−5239.

[29] Orlandi, C., Jacques, C., Duplan, H., Debrauwer, L., Jamin, E. Miniaturized two-dimensional heart cutting for LC–MS-based metabolomics. *Anal.Chem.* 2023, *95*, 2822−2831.

[30] Zhu, P., Dubbelman, A., Hunter, C., Genangeli, M., Karu, N., Harms, A., Hankemeier, T. Development of an untargeted LC–MS metabolomics method with postcolumn infusion for matrix effect monitoring in plasma and feces. *J. Am. Soc. Mass Spectrom.* 2024 [https://doi.org/10.1021/jasms.3c00418].

[31] Phapale, P., Rai, V., Mohanty, A., Srivastava, S. Untargeted metabolomics workshop report: Quality control considerations from sample preparation to data analysis. *J. Am. Soc. Mass Spectrom.* 2020, *31*, 2006−2010.

[32] Ivanova, B., Spiteller, M. A mass spectrometric stochastic dynamic diffusion approach to selective quantitative and 3D structural analyses of native cyclodextrins by electrospray ionization and atmospheric pressure chemical ionization methods. *Bioorg. Chem.* 2019, *93*, 103308.

[33] Ivanova, B., Spiteller, M. Stochastic dynamic electrospray ionization mass spectrometric quantitative analysis of metronidazole in human urine. *Anal. Chem. Lett.* 2022, *12*, 322–348.

[34] Ivanova, B. Stochastic dynamic mass spectrometric quantitative and structural analyses of pharmaceutics and biocides in biota and sewage sludge. *Int. J. Mol. Sci.* 2023, *24*, 6306.

[35] Ivanova, B., Spiteller, M. Stochastic dynamics mass spectrometric 3D structural analysis of N-glycans of fetal bovine serum — an Experimental and theoretical study. *Med. Sci. Forum* 2022, *14*, 2.

[36] Ivanova, B. 3D structural analysis of alanyl-containing peptides of silac-proteomics — an Approach to stochastic dynamics. *SSRN eJ.* 2023 [https://dx.doi.org/10.2139/ssrn.4606434].

[37] Upadyshev, M., Ivanova, B., Motyleva, S. Mass spectrometric identification of metabolites after magnetic-pulse treatment of infected *Pyrus communis* L. microplants. *Int. J. Mol. Sci.* 2023, *24*, 1677.

[38] Frisch, M., Trucks, G., Schlegel, H., Scuseria, G., Robb, M., Cheeseman, J., Fox. D. Gaussian, Inc., Pittsburgh, Wallingford CT.

[39] Helgaker, T., Jensen, H., Jrgensen, P., Olsen, J., Ruud, K., Agren, H., Auer, A., Bak, K., Bakken, V., Christiansen, O., Vahtras, O. Dalton Program Package, 2011; [http://www.daltonprogram.org/download.html].

[40] Gordon, M., Schmidt, M. Advances in electronic structure theory: GAMESS a decade later, In *Theory and applications of computational chemistry: The first forty years*. Dykstra, C., Frenking, G., Kim, K., Scuseria, G. eds. Elsevier, Amsterdam, 2005; pp. 1167–1189.

[41] Nielsen, A., Holder, A. Gauss View 3.0, User's Reference. GAUSSIAN Inc., Pittsburgh GausView03 Program Package, 2009; [www.gaussian.com/g_prod/gv5.htm].

[42] Lee, J., Lin, L., Head-Gordon, M. Systematically improvable tensor hypercontraction: interpolative separable density-fitting for molecules applied to exact exchange, second- and third-order Møller–Plesset perturbation theory. *J. Chem. Theory Comput.* 2020, *16*, 243–263.

[43] Zhao, Y., Truhlar, D. Density functionals with broad applicability in chemistry. *Accs Chem. Res.* 2008, *41*, 157–167.

[44] Zhao, Y., Truhlar, D. The M06 suite of density functionals for main group thermochemistry, thermochemical kinetics, noncovalent interactions, excited states, and transition elements: two new functionals and systematic testing of four M06-class functionals and 12 other functionals. *Theor. Chem. Accts.* 2008, *120*, 215–241.

[45] Ni, Z., Wang, Y., Li, W., Pulay, P., Li, S. Analytical energy gradients for the cluster-in-molecule MP2 method and its application to geometry optimizations of large systems. *J. Chem. Theory Comput.* 2019, *15*, 3623–3634.^t.

[46] Schira, R., Latouche, C. FT *vs*. TDDFT *vs*. TDA to simulate phosphorescence spectra of Pt- and Ir-based complexes. *Dalton Trans.* 2021, *50*, 746–753.

[47] Tamm, I. Relativistic interaction of elementary particles. *J. Phys. (URSS)* 1945, *9*, 449–460.

[48] Dancoff, S. Non-adiabatic meson theory of nuclear forces. *Phys. Rev.* 1950, *78*, 382–385.

[49] Shu, Y., Levine, B. Do excited silicon-oxygen double bonds emit Light? *J. Phys. Chem. C* 2014, *118*, 7669–7677.

[50] Burkert, U., Allinger, N. Molecular Mechanics. In *ACS Monograph* 177, American Chemical Society, Washington D.C. 1982; pp. 1–339.

[51] Allinger, L. Conformational analysis. 130. MM2. A hydrocarbon force field utilizing V1 and V2 torsional terms. *J. Am. Chem. Soc.* 1977, *99*, 8127–8134.

[52] Boys, S., Bernardi, F. The calculation of small molecular interactions by the differences of separate total energies. Some procedures with reduced errors. *Mol. Phys*. 1970, *19*, 553–555.

[53] Simon, S., Duran, M., Dannenberg, J. How does basis set superposition error change the potential surfaces for hydrogen-bonded dimers? *J. Chem. Phys*. 1996, *105*, 11024–11031.

[54] Polack, E., Dusson, G., Stamm, B., Lipparini, F. Grassmann extrapolation of density matrices for Born–Oppenheimer molecular dynamics. *J. Chem. Theory Comput*. 2021, *17*, 6965–6973.

[55] Plasser, F., Crespo-Otero, R., Pederzoli, M., Pittner, J., Lischka, H., Barbatti, M. Surface hopping dynamics with correlated single-reference methods: 9H-Adenine as a case study. J. Chem. Theory Comput. 2014, *10*, 1395–1405.

[56] Plasser, F., Gomez, S., Menger, M., Mai, S., Gonzalez, L. Highly efficient surface hopping dynamics using a linear vibronic coupling model. Phys. Chem. Chem. Phys. 2019, *21*, 57–69.

[57] Tully, J. Molecular dynamics with electronic transitions. *J. Chern. Phys*. 1990, *93*, 1061–1071.

[58] Hirata, S., Head-Gordon, M. Time-dependent density functional theory within the Tamm–Dancoff approximation. *Chem. Phys. Lett.* 1999, *314*, 291–299.

[59] Zhao, X., Merritt, I., Lei, R., Shu, Y., Jacquemin, D., Zhang, L., Xu, X., Vacher, M., Truhlar, D. Nonadiabatic coupling in trajectory surface hopping: accurate time derivative couplings by the curvature-driven approximation. *J. Chem. Theory Comput*. 2023, *19*, 6577–6588.

[60] Ryde, U. How many conformations need to be sampled to obtain converged QM/MM energies? The curse of exponential averaging. *J. Chem. Theory Comput*. 2017, *13*, 5745–5752.

[61] Douroudgari, H., Vahedpour, M. A computer-aided method for controlling chemical resistance of drugs using RRKM theory in the liquid phase. *Sci. Rep.* 2021, *11*, 22971.

[62] Cangtao, Y., Ran, G., Jiulin, D. The rate coefficients of unimolecular reactions in the systems with power-law distributions. *ArXiv* 2014; [https://doi.org/10.48550/arXiv.1404.7213] (Accessible for 29.04.2014.)

[63] Mishra, P., Kumar, A. Molecular electrostatic potentials and fields: Hydrogen bonding, recognition, reactivity, and modelling; In Molecular electrostatic potentials; Concepts and Applications; Theoretical and Computational Chemistry, Vol. 3; Elsevier Sci. B.V. Amsterdam; 1996, pp. 257–296.

[64] Zhu, Q., Wu, Y., Zhao, S., Cieplak, P., Duan, Y., Luo, R. Streamlining and optimizing strategies of electrostatic parameterization. *J. Chem. Theory Comput*. 2023, *19*, 6353–6365.

[65] Weinhold, F., Landis, C. Discovering chemistry with natural bond orbitals. John Wiley & Sons, Hoboken, NJ; 2012 (Sections 4 and 5.)

[66] Jiao, Y., Weinhold, F. NBO/NRT Two-state theory of bond-shift spectral excitation. *Molecules* 2020, *25*, 4052.

[67] Demissie, T. Relativistic effects on the NMR parameters of Si, Ge, Sn, and Pb alkynyl compounds: Scalar *versus* spin–orbit effects. *J. Chem. Phys*. 2017, *147*, 174301.

[68] Jakubowska, K., Pecul, M., Ruud, K. Relativistic four-component DFT calculations of vibrational frequencies. *J. Phys. Chem. A* 2021, *125*, 10315–10320.

[69] Malkin, E., Malkin, I., Malkina, O., Malkin, V., Kaupp, M. Scalar relativistic calculations of hyperfine coupling tensors using the Douglas-Kroll-Hess method with a finite-size nucleus model. *Phys. Chem. Chem. Phys*. 2006, *2006*, 4079–4085.

[70] Stevens, W., Krauss, M. *Ab initio* effective spin–orbit operators for use in atomic and molecular structure calculations. Results for carbon and silicon. *Chem. Phys. Lett.* 1982, *86*, 320–324.

[71] Kupka, T., Stachow, M., Chelmecka, E., Pasterny, K., Stobinska, M., Stobinski, L., Kaminsky, J. Efficient modeling of NMR parameters in carbon nanosystems. *J. Chem. Theory Comput.* 2013, *9*, 4275–4286.

[72] Kapusta, K., Voronkov, E., Okovytyy, S., Korobov, V., Leszczynski, J. Reconstruction of STO-3G family basis set for the accurate calculation of magnetic properties. *Russian J. Phys. Chem. A* 2018, *92*, 2827–2834.

[73] Kwon, H., Curtin, G., Morrow, Z., Kelley, C., Jakubikova, E. Adaptive basis sets for practical quantum computing. *Int. J. Quantum. Chem.* 2023, *123*, e27123.

[74] Hehre, W., Stewart, R., Pople, J. Self-consistent molecular-orbital methods. I. Use of Gaussian expansions of Slater-type atomic orbitals. *J. Chem. Phys.* 1969, *51*, 2657.

[75] Hong, C., Tsai, T., Chou, J., Chen, P., Tsai, P., Chen, Y., Kuo, E., Srolovitz, D., Hu, A., Cheng, Y., Goan, H. Accurate and efficient quantum computations of molecular properties using daubechies wavelet molecular orbitals: A benchmark study against experimental data. *ArXiv* 2022, [arXiv:2205.14476v1].

[76] Popov, A., Obkhodsky, A., Kuznetsov, S. Calculation method of materials'properties on the basis of first-principles methods (or *ab initio*) for quantum-chemical modeling program package ANTARES. *J. Indu. Pollut. Contr.* 2016, *32*, 418–423.

[77] Burton, H. Energy landscape of state-specific electronic structure theory. *J. Chem. Theory Comput.* 2022, *18*, 1512–1526.

[78] Shajan, A., Manathunga, M., Goetz, A., Merz, Jr., K. Geometry optimization: A comparison of different open-source geometry optimizers. *J. Chem. Theory Comput.* 2023, *19*, 7533–7541.

[79] Sayfutyarova, E. Molecular □-orbital construction for non-planar conjugated systems. *J. Chem. Theory Comput.* 2024, [https://doi.org/10.1021/acs.jctc.3c00949].

[80] Lapidus, A., Gyul'maliev, A., Zhagfarov, F., Yarkova, T. Quantum-chemical evaluation of the energies of solvation of humic substances in various solvents, *Solid Fuel Chem*. 2022, *56*, 342–346.

[81] Tomer, H., Goswami, B., Modak, P., Alam, M., Ahmad, S., Antony, B. Low-energy electron scattering from pyrrole and its isomers. *J. Phys. Chem. A* 2023, *127*, 10464–10480.

[82] Kupka, T., Mnich, A., Broda, M. Performance of revised STO(1M)-3G basis set for prediction of 5-fluorocytosine chemical shifts. *Magn Reson Chem*. 2019, *57*, 489–498.

[83] Lim, H., Jeon, H., Rhee, J., Oh, B., No, K. Quantum computational study of chloride attack on chloromethane for chemical accuracy and quantum noise effects with UCCSD and k-UpCCGSD ansatzes. *Sci. Rep.* 2022, *12*, 7495.

[84] Kulik, H., Seelam, N., Mar, B., Martínez, T. Adapting DFT+U for the chemically motivated correction of minimal basis set incompleteness. *J. Phys. Chem. A* 2016, *120*, 5939–5949.

[85] Kromann, J., Welford, A., Christensen, A., Jensen, J. Random *versus* systematic errors in reaction enthalpies computed using semiempirical and minimal basis set methods. *ACS Omega* 2018, *3*, 4372–4377.

[86] Pohl, M., Schmitt, M., Kleinermanns, I. Vibrational spectroscopy of size-assigned *p*-cresol/H,O-clusters in the S_0 and S_1 state. *Chem. Phys. Lett.* 1991, *177*, 252–260.

[87] Khamis, T., Diab, A., Zahra, M., El-Dahmy, S., Al-Hameed, B., Abdelkhalek, A., Said, M., Abdellatif, H., Fericean, L., Banatean-Dunea, I., Arisha, A., Attia, M. The antiproliferative activity of *Adiantum pedatum* extract and/or piceatannol in phenylhydrazine-Induced colon cancer in male albino rats: The miR-145 expression of the PI-3K/Akt/p53 and Oct4/Sox2/Nanog pathways. *Molecules* 2023, *28*, 5543.

[88] Robbins, R. ; Robbins, R. Phenolic acids in foods: An overview of analytical methodology. *J. Agric. Food Chem.* 2003, *51*, 2866–2887.

[89] Collins, D., Jones, E. Regioselective 2,3-Di-*O*-silylation of methyl 6-chloro-6-deoxy-α-D-glucopyranoside. *Austr. J. Chem.* 2010, *47*, 1419–1421.

[90] Habib, S., Larnaud, F., Pfund, E., Barragán, T., Lequeux, T., Mellet, C., Goekjian, P., Gueyrard, D. Synthesis of substituted exo-glucals *via* a modified Julia olefination and identification as selective β-glucosidase inhibitors. *Org. Biomol. Chem.* 2014, *12*, 690–699.

[91] Lapierre, C., Voxeur, A., Karlen, S., Helm, R., Ralph, R. Evaluation of feruloylated and p-coumaroylated arabinosyl units in grass *Arabinoxylans* by acidolysis in dioxane/methanol. *J. Agric. Food Chem.* 2018, *66*, 5418–5424.

[92] Koizumi, K., Utamura, T. Trityl derivatives of cellobiose. I. Synthesis and isolation of 6- and 6'-O-tritylcellobioses. *Chem. Pharm. Bull.* 1981, *29*, 2776–2784.

[93] Marinova, G., Batchvarov, V. Evaluation of the methods for determination of the free radical scavenging activity by DPPH. *Bulgarian J. Agric. Sci.* 2011, *17*, 11–24.

[94] Han, X., Yang, K., Gross, R. Multi-dimensional mass spectrometry-based shotgun lipidomics and novel strategies for lipidomic analyses. *Mass Spectrom. Rev.* 2012, *31*, 134–178.

[95] Bergman, N., Shenchenko, D., Bergquist, J. Approaches for the analysis of low molecular weight ompounds with laser desorption/ionization techniques and mass spectrometry. *Anal. Bioanal. Chem.* 2014, *406*, 49–61.

[96] Kelley, C. Iterative Methods for optimization. *SIAM Front. Appl. Math.* 2009, *18*, 1.

[97] Otto, M. *Chemometrics*. 3rd ed. Wiley, Weinheim, 2017; pp. 1–383.

[98] Apache OpenOffice, [http://de.openoffice.org].

[99] Madsen, K., Nielsen, H., Tingleff, T. *Informatics and mathematical modelling*. 2nd ed. DTU Press, 2004.

[100] Miller, J., Miller, J. *Statistics and chemometrics for analytical chemistry*. Pentice Hall, London, 1988; pp. 1–271.

[101] Taylor, J. *Quality assurance of chemical measurements*. Lewis Publishers Inc. Michigan, 1987; pp. 1–328.

[102] Schroee, G., Trenkler, D. Exact and randomization distributions of Kolmogorov–Smirnov tests two or three samples. *Comput. Stat. Data Anal.* 1995, *20*, 185–202.
[103] Fay, M., Proschan, M. Wilcoxon–Mann–Whitney or t-test? On assumptions for hypothesis tests and multiple interpretations of decision rules. *Stat. Surv.* 2010, *4*, 1–39.
[104] Freidlin, B., Gastwir, J. Should the median test be retired from general use? *Am. Stat.* 2000, *54*, 161–164.
[105] Xiao, Z. et al. Earthworms affect plant growth and resistance against herbivores: A meta-analysis. *Functional* Ecology. 2018, 32.1, 150–160.
[106] Shi, J., Zhao, J., Zhang, Y., Wang, Y., Tan, C., Xu, Y., Liu, Y. Windows scanning multiomics: Integrated metabolomics and proteomics. *Anal. Chem.* 2023 [https://doi.org/10.1021/acs.analchem.3c03785].
[107] Peterson, A., Balloon, A., Westphall, M., Coon, J. Development of a GC/quadrupole-Orbitrap mass spectrometer, Part II: New approaches for discovery metabolomics. *Anal. Chem.* 2014, *86*, 10044–10051.

Chapter 2

Advances in Mass Spectrometry-Based Proteomics: Workflow and Recent Biological Applications

Thi Phuong Truc Nguyen[1,2]
Thi Quynh Nhu Nguyen[1,2]
Rajat Ujjainiya[4,5]
Shantanu Sengupta[4,5]
and Thi Thu Hoai Nguyen[1,2,3,*]
[1]School of Biotechnology, International University, Ho Chi Minh City, Vietnam
[2]Vietnam National University Ho Chi Minh City, Ho Chi Minh City, Vietnam
[3]Research Center for Infectious Diseases, International University, Ho Chi Minh City, Vietnam
[4]Cardiorespiratory Disease Biology Unit, Institute of Genomics and Integrative Biology, New Delhi, India
[5]Council of Scientific and Industrial Research, New Delhi, India

Abstract

Mass Spectrometry (MS)-based proteomics is essential for understanding complex biological systems by allowing comprehensive analysis of protein composition and interactions. This chapter presents a comprehensive workflow for MS-based proteomics, along with a review of recent innovations and the integration of bioinformatics tools to improve peptide identification and quantification. It highlights the essential role of bioinformatics platforms in handling large-scale

---

* Corresponding Author's Email: ntthoai@hcmiu.edu.vn

In: Advances in Biology. Volume 9
Editor: Charles D. Grant
ISBN: 979-8-89530-574-4

datasets, optimizing peptide identification, and supporting quantitative analysis. The chapter also covers applications across diverse fields, including disease biomarker discovery, drug development, and molecular pathway analysis. By facilitating deeper insights into protein interactions and modifications, the MS-based proteomics approaches discussed in this chapter offer new perspectives for understanding pathogenesis and enable more accurate modeling of biological pathways, advancing our knowledge of disease mechanisms. Furthermore, recent developments in MS-based proteomics that address key challenges in sensitivity, data analysis, and proteome coverage are highilighted. Significant advancements include innovations in single-cell proteomics and cutting-edge technologies like nanodroplet sample preparation and machine learning algorithms. These improvements have substantially enhanced the sensitivity and precision of proteome analysis, allowing researchers to explore cellular heterogeneity and detect subtle changes in protein expression. Collectively, these proteomic advancements not only refine existing methodologies but also broaden the potential for transformative discoveries in biomedical research, ultimately contributing to the development of novel diagnostic and therapeutic strategies.

Keywords: advanced technologies, biomedical, biomarker discovery, drug development, mass spectrometry, proteomics

Introduction

In 1994, an Australian scientist named Marc Wilkins first used the term "proteome" in a conference at Siena, Italy (Beeton-Kempen, 2010). The phrase serves as a term for indicating all the proteins that a cell, an organism or a species is capable of producing (Ali-Khan et al., 2003). Every species has a distinct proteome of its own. Moreover, the proteome of a cell is the cell's particular proteins being expressed in a certain physiological context at a specific time. Thus, when scientists mention the proteome, they may also mean the proteome with respect to a specific condition such as the proteome of a newborn versus that of a mature organism or the proteome of a defined cell type or tissue within the body. (Beeton-Kempen, 2010)

The study of proteome, or proteomics, focuses on which proteins are present in a sample and how they interact with one another. This will further enhance our understanding of the functions they perform in a cell or an organism (Beeton-Kempen, 2010). At the beginning, there was some confusion over what proteomics actually encompassed, since some researchers considered proteomics to be merely a new name for the old field – protein biochemistry. However, instead of emphasizing on a single protein or simple macromolecular complexes as in most prior protein biochemical studies, proteomics provides a broader, more comprehensive, aggressive, and systematic approach to the research of biological systems (Ali-Khan et al., 2003). Furthermore, others also stated that mRNA, the intermediate between genes and proteins, could be analyzed to deduce protein expression (Beeton-Kempen, 2010). Though, the study of mRNA levels did not account for protein post-translational modifications, complex formation or even localization, which are essential to identify the proteins and understand their functions (Beeton-Kempen, 2010). Therefore, proteome analysis deserves to become a distinct discipline with a variety of approaches to the biological system.

Although numerous technological advancements have influenced current proteomics studies, two crucial developments aided in the large-scale, systematic analyses of proteins (Ali-Khan et al., 2003). First, back in the 1970s, proteins were separated using 2-D gel electrophoresis. This method continues to be a key tool for comparing protein levels quantitatively between two or more samples. However, because of the lack of effective technology to identify and further characterize the separated proteins at the time, the analytical bound of the early 2-D gel studies was rather constrained. Then, about 20 years later, the advent of mass spectrometry (MS) led to the feasibility of rapid, high-throughput identification of multiple proteins per sample (Ali-Khan et al., 2003; Beeton-Kempen, 2010). John Fenn and Koichi Tanaka received the 2002 Nobel Prize in Chemistry for developing two ionization techniques that revolutionized proteomics in general and biomedical mass spectrometry in particular.

MS is an analytical technique whereby chemical or biological substances are determined by separating ions in electric and magnetic fields according to their mass ratios, or it can be said that MS measures the mass-to-charge ratio of ions (Han et al., 2008). In proteomics, mass provides characteristics regarding the structure, chemical modification, and identity of the protein (C. E. Parker et al., 2010). The use of MS has widespread applications and has become an indispensable tool for proteomics research (Han et al., 2008).

There are three key components of a mass spectrometer: a source, an analyzer, and a detector. The source takes responsibility for converting analyte molecules into ions. It provides energy to transform peptides from liquid phase to gas phase, and subsequently ionizes all the peptides. Two major ionization technologies include matrix-assisted laser desorption/ionization (MALDI) and electron spray ionization (ESI). The analyzer, such as time-of-flight (TOF) or ion trap separates different ionized peptides according to their mass-to-charge (m/z) values. The detector finally captures these charged particles and records the number of each m/z value, which results in the mass spectra with the y-axis as relative abundance and the x-axis as m/z values (Beeton-Kempen, 2010; Han et al., 2008; C. E. Parker et al., 2010).

For MS-based bottom-up proteomics, the intact protein sample has to be processed before being analyzed in a mass spectrometer. Complex macromolecules such as proteins are almost difficult to be thoroughly analyzed. Therefore, the digestion step needs to be performed to cleave the protein into several smaller fragments called peptides. For digestion, specific proteases such as LysC and trypsin are commonly used in order to generate peptides ranging from 7 to 30 amino acids in length which are most appropriate for analysis. These digested peptides are then analyzed in LC-MS which is the combination of liquid chromatography (LC) with mass spectrometry (MS). The commonly used LC for peptide analysis is reversed phase HPLC with the column packed with hydrophobic stationary phase via which peptides are partially separated based on theirhydrophobicity. After that, the separated peptides are subjected to the mass spectrometer for further analysis (Shuken, 2023).

For MS-based top-down proteomics, it works with intact protein ions without a digestion step. After separating proteins from the mixture, the intact proteins are directly ionized and detected via mass spectrometer to acquire data for analysis. Despite the high percentage coverage of protein sequences which enhances the ability of localizing and characterizing post-translational modifications, this method is not commonly used due to its high cost as well as the difficulty of handling proteins which are more complicated than small peptides.

Tandem mass spectrometry (MS/MS) can be performed on both top-down proteomics and bottom-up proteomics. This is a technique where two mass analyzers are coupled together to analyze complex mixtures and involve two stages of MS. In the first stage, it measures m/z values and intensities of precursor ionized peptides, which are displayed in the MS1 spectrum where relative quantification can be performed. However, it lacks the necessary information for naming the peptide identities/sequences. Therefore, one or more detected ion peptides are then selected and fragmented, mostly by collision with gas. The instrument subsequently analyzes the fragments and produces MS2 spectrum which can provide information for both protein identification as well as quantification (Shuken, 2023).

This chapter focuses on MS-based bottom-up proteomics and its practical applications in life sciences.

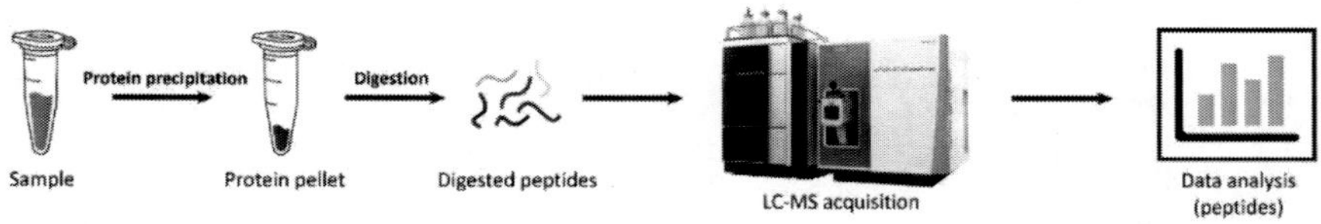

Figure 1. Basic bottom-up proteomics workflow. Figure is generated using Biorender free version 2024.

Sample Preparation for Proteomic Analysis

Sample preparation is the most critical step in MS-based proteomics, as it directly impacts the quality and reliability of the results obtained. While it is essential to acknowledge the uniqueness of each proteomics sample, there are some universal principles that can be applied to all samples. To obtain optimal results, careful consideration must be given to the entire process, starting from the initial sample preparation, through LC-MS/MS acquisition, to the subsequent bioinformatics analyses. It is advisable to focus on systematic prevention of contaminants rather than relying solely on their reduction in retrospect, although achieving complete elimination may not always be feasible depending on the experimental setup.

The sample preparation steps for bottom-up proteomics as described in Figure 2.

Protein Extraction and Purification

The first step for most proteomics- analysis methods (e.g., Western Blot, SDS-PAGE, MS) is protein extraction and purification. The goal of protein extraction is to enhance the accuracy and depth of information gained through MS analysis. After the collection of biological samples (e.g., tissues, cells, bodily fluids), they are processed through homogenization, cell lysis, and the removal of interfering substances to ensure their integrity and preservation. Proteins from the samples are then extracted through precipitation using acetone or trichloroacetic acid (Wang et al., 2018).

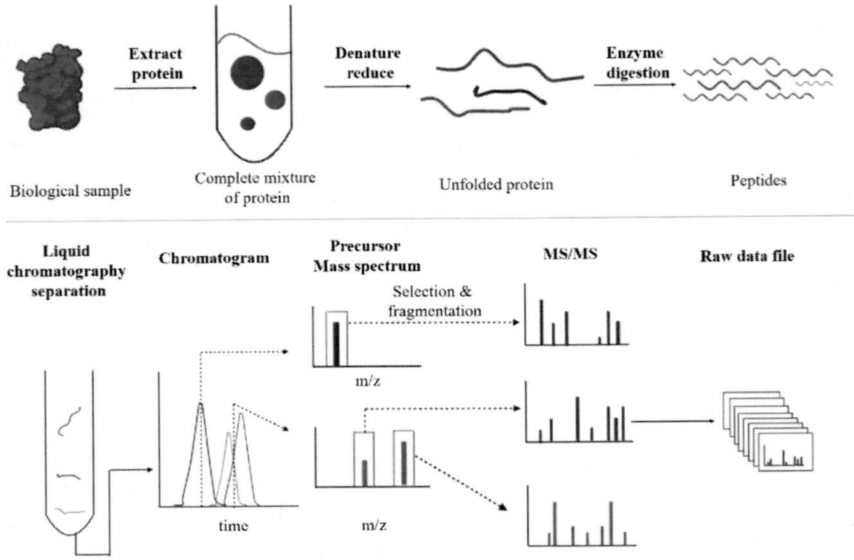

Figure 2. Sample preparation for mass spectrometry analysis. Protein is extracted, denatured, and digested into peptides, followed by separation using liquid chromatography. As peptides elute, they are ionized and introduced into the mass spectrometer, where the mass-to-charge ratio (m/z) of precursor ions is measured. These ions are isolated and fragmented, and the resulting fragment masses are analyzed to identify peptides through database searches or computational methods. Figure is generated using Biorender free version 2024.

Following precipitation, protein pellets are solubilized in suitable buffers such as Tris Urea, Ammonium bicarbonate, and others. Purified proteins are denatured (unfolded) with urea or sodium deoxycholate (DOC). Denaturation allows for selective accessibility to subsequent reagents and proteases. Enhancing the process of denaturation may lead to an improvement in proteolytic efficiency, hence resulting in increased repeatability of the sample preparation (Proc et al., 2010). Often, the protein concentration in the sample is measured using the bicinchoninic acid (BCA) (Smith et al., 1985) or Bradford assay (Kielkopf et al., 2020).

In order to disrupt the covalent disulfide bonds that contribute to a protein's tertiary structure and to prevent the reformation of disulfide bonds, proteins are further subjected to reduction and alkylation processes. This involves the reduction of proteins and afterward capping the resulting free cysteines. The process of reducing cysteine residues in a protein enhances the accessibility for proteolysis and streamlines the search process in databases by removing any peptide complexes that are connected via disulfide bonds (Sechi

& Chait, 1998). For this process, several reagents have been used, such as dithiothreitol (DTT) or tris-(2-carboxyethyl)-phosphine (TCEP). These reducing agents break disulfide bonds (S-S) within proteins, converting them into free sulfhydryl (SH) groups. This is followed by the introduction of an alkylating agent like iodoacetamide (IAA) to the mixture. Alkylation blocks the free cysteine residues, preventing the reformation of disulfide bonds during subsequent steps. Alkylation is typically performed in the dark to avoid light-induced degradation of reagents and unwanted side reactions (Crankshaw & Grant, 2001; Gundry et al., 2010; Hill et al., 2009).

Protein Digestion

Proteins can be digested into peptides using three general approaches: in-solution digestion (Medzihradszky, 2005), in-gel digestion (Shevchenko et al., 2007), and filter-assisted sample preparation (Wiśniewski, 2018).

In-Solution Digestion

In-solution digestion is a common and well-established technique because of its simplified procedure and scalability for analyzing protein samples (from less than 10 μg to larger than 1 mg). This method involves denaturation of proteins to disrupt their native structures and facilitate solubilization as discussed. After the reduction and alkylation of disulfides, it is important to remove any contaminants, including salts, reducing and alkylating reagents, detergents, and other small molecule that may be present in the matrix prior to digestion. This step is crucial to lower the concentration of denaturants (urea or guanidine) to levels that are compatible with the protease's activity. A specific protease, such as trypsin, is added to catalyze the proteolytic cleavage. The enzyme-to-substrate ratio, or the amount of protease relative to the protein in the sample, is a critical parameter that affects digestion efficiency and peptide yield (Zhang et al., 2017). The trypsin-treated sample is then incubated at an appropriate temperature (e.g., 37°C) for an extended period, typically ranging from several hours to overnight (12-16 hours). The incubation allows the protease to cleave peptide bonds at specific sites within the proteins (Goodman et al., 2018). Trypsin cleaves proteins specifically at the C-terminus of lysine (K) and arginine (R) residues, generating peptides with basic residues at one end. These peptides are well-suited for subsequent MS analysis (Olsen et al., 2004). To stop the digestion activity, an acid, such as formic acid or trifluoroacetic acid (TFA), is added to the reaction mixture.

This acidification serves multiple purposes, including the inactivation of the protease and the preparation of peptides for MS analysis. Acidification also helps to minimize sample contamination and preserve peptide stability (Doellinger et al., 2020; Zhang et al., 2017). Following digestion, peptides undergo desalting that removes salts, detergents, and other interfering substances from the mixture before analysis. This could be achieved using off-line methods, such as employing reverse phase solid phase extraction cartridges or tips (Karch et al., 2016). Alternatively, on-line desalting can be performed by utilizing a trap column prior to conducting MS analysis (Yoshida et al., 2006).

In-Gel Digestion

For in-gel digestion, proteins are first separated by gel electrophoresis and visualized by gel stains such as Coomassie Blue, Colloidal Coomassie, and glutaraldehyde-free silver stain. The gel bands that contain the protein(s) of interest are then excised, destained, reduced, and alkylated in order to enhance the digestion and extraction of peptides (Gundry et al., 2009).

After reduction and alkylation, the gel bands undergo digestion with a protease. LysC is a protease that specifically cleaves peptide bonds after lysine (K) residues. Its cleavage specificity results in peptides with lysine residues at their N-terminus while trypsin cleaves the peptide bond at C-terminal of lysine (K) and arginine (R) (Yen et al., 2006) (Glatter et al., 2012). Combination of LysC and trypsin for protein digestion is preferred because of its general efficiency, effectiveness and its pronounced preference for cleaving after K and R residues, yielding peptides with optimal lengths for MS analysis. The peptides thus produced are expected to carry positive charges, as K and R, being primary cleavage sites, possess high pKa values.

Once digestion is complete, the next step is to extract the peptides from the gel piece. This is typically done by adding an extraction solution that helps to release the peptides from the gel matrix. Common extraction solutions include acetonitrile, trifluoroacetic acid (TFA), or a mixture of organic solvents and acids (Greco et al., 2012; Paulo, 2016). The choice of extraction solution depends on the specific analytical techniques and conditions used. The gel piece is often agitated or sonicated in the extraction solution to facilitate the release of peptides from the gel. During this process, the peptides are solubilized and released into the extraction solution. The mixture is gently agitated or sonicated to aid in the extraction process (Shevchenko et al., 2007).

The extracted peptides may contain salts and other impurities from the gel and the extraction process. Desalting is performed to remove these contaminants and prepare the peptides for MS analysis. Common desalting methods include using C18 ZipTips (Nika et al., 2013), solid-phase extraction (SPE) cartridges (Badawy et al., 2022), or other chromatographic techniques to purify and concentrate the peptides. Following this, the desalted peptides are subjected to drying to eliminate any residual solvent, yielding a concentrated sample. This can be achieved by employing a vacuum concentrator or by air-drying under carefully controlled conditions to prevent peptide degradation (Rosenfeld et al., 1992). Subsequently, the dried peptides are eventually reconstituted in a compatible aqueous buffer LC-MS/MS analysis.

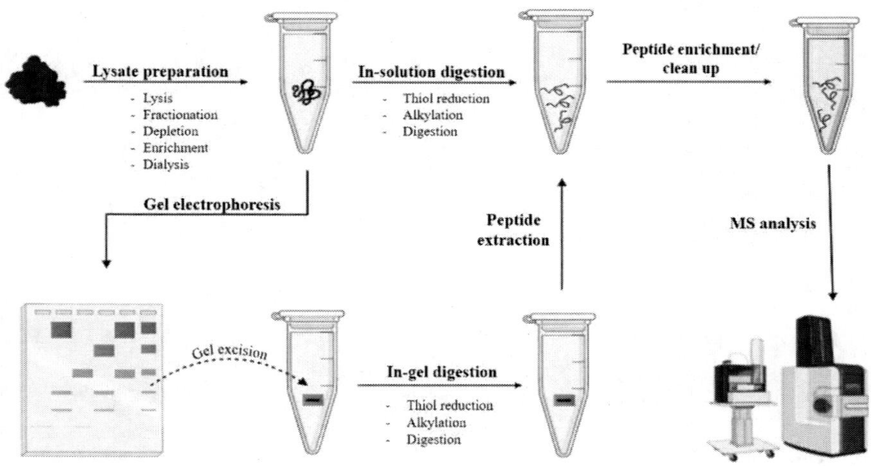

Figure 3. In-solution and in-gel digestion workflow. Figure is generated using Biorender free version 2024.

Filter-Aided Sample Preparation (FASP)

Filter aided sample preparation (FASP) is a proteomic approach that integrates the advantages of both in-gel and in-solution digestion. This method involves the solubilization of the proteins using sodium dodecyl sulfate (SDS), which is then replaced by urea using a conventional filtering apparatus. The workflow of FASP consists of three critical steps: (i) elimination of SDS, (ii) chemical modification of thiol-moieties, and (iii) digestion of proteins followed by the purification of peptides (Wiśniewski, 2018).

In the first stage of FASP, SDS extract is diluted using a buffer solution that contains 8M urea. A high concentration of urea is essential for both the preservation of proteins in solution and for shrinking and dissociation of detergent micelles, which is crucial for the efficient and complete removal of the detergent. This buffer, along with the detergent and other small molecules present in the sample, is eliminated by a series of filtering steps facilitated by centrifugation. In the second stage, denatured proteins undergo chemical modifications, mostly, cysteine residues undergo carbamidomethylation with the use of iodoacetamide. Typically, a duration of 15-20 minutes is necessary for a single filtration step to occur at room temperature. The alkylation process of the thiol groups in cysteine is performed following the first filtration and persists throughout the subsequent filtration stages until the elimination of the alkylating agent. Subsequently, the process of ultrafiltration is used to facilitate the exchange of urea, followed by the treatment of proteins with a protease (Wiśniewski, 2019).

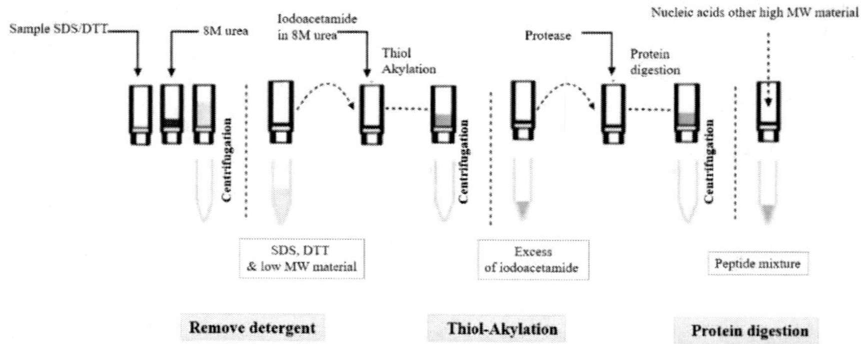

Figure 4. The filter-aided sample preparation workflow. Figure is generated using Biorender free version 2024.

Protein digestion, as it permits the activity, can occur either with or without urea. However, it is advisable to perform digestion in the presence of urea when dealing with samples primarily composed of water-insoluble proteins or containing endogenous proteases, such as lysates from mouse intestinal mucosa (Wiśniewski et al., 2014). When employing denaturants, it is crucial to consider that many proteases experience a loss of enzymatic activity and specificity under denaturing conditions. To overcome this limitation, digestion in the absence of urea is recommended. Among the proteases that are evaluated, cleavage with trypsin, endoproteinases LysC, and ArgC consistently yield the highest peptide identification rates during LC-MS/MS analysis (Wiśniewski et al., 2014; Wiśniewski et al., 2011; Zielinska et al., 2009). After the digestion step, peptides liberated from the proteins are gathered using ultrafiltration. This process facilitates the preservation of undigested protein fragments and other native macromolecules found in the lysate, such as nucleic acids and oligosaccharides. Ensuring the purity of the protein digest is essential for achieving high-resolution liquid chromatography of the peptides, maintaining run-to-run reproducibility, and extending the lifespan of the chromatography column (Wiśniewski, 2018).

Peptide Separation Using Liquid Chromatography (LC)

Following protein digestion and sample clean-up steps, the peptides are introduced into liquid chromatography (LC) column before the mass spectrometer run. This phase serves as a critical bridge between sample preparation and mass spectrometric analysis, aiming to resolve complex peptide mixtures generated through the enzymatic digestion of proteins. LC employs chromatographic columns, often C18-based, where peptides are introduced and separated based on their hydrophobicity. The process begins with the introduction of peptides onto the column, followed by their progressive elution utilizing a solvent gradient. This modulation of solvent composition results in a gradual and systematic release of peptides from the column, unraveling the chromatogram with distinct peaks, each representing a unique peptide. These resolved peaks, arising from the differential retention and elution times of peptides, embody the foundation upon which subsequent MS analysis is built. By meticulously partitioning and resolving the peptide constituents, LC remarkably diminishes sample complexity, bestows heightened detection sensitivity, and enriches the quality of ensuing MS data. In LC gradient elution, water and acetonitrile serve as common mobile phase solvents, particularly in reverse-phase chromatography methods. The ratio of water to acetonitrile is carefully adjusted throughout the gradient to optimize the separation of peptides. (D'Cunha et al., 2018; Dembek & Bocian, 2020).

Mass Spectrometry Analysis

MS involves a series of intricate steps, including ionization, followed by mass analysis, and culminating in fragmentation for accurate peptide identification. Ionization marks the initial phase, where peptides, generated through enzymatic digestion of proteins, are converted into gas-phase ions suitable for analysis within the mass spectrometer. Common ionization techniques include

Electrospray Ionization (ESI)(Fenn et al., 1989) and Matrix-Assisted Laser Desorption/Ionization (MALDI)(Karas & Hillenkamp, 2002). ESI involves the nebulization of peptide solutions, generating charged droplets that lose solvent to yield gas-phase ions, while MALDI uses a matrix to assist in ionization upon laser irradiation, producing ions from the sample. This critical ionization step sets the stage for subsequent mass analysis and downstream MS/MS, enabling comprehensive peptide identification and deciphering the intricate proteomic landscape.

After the ionization, analytes proceed through the mass analyzers. Various analyzers like Time-of-Flight (TOF)(Ogorzalek Loo et al., 2001), Quadrupole(Yost & Boyd, 1990), Ion Trap(Chi et al., 2007), or Orbitrap(Makarov, 2000) (LTQ-Orbitrap(Makarov et al., 2006)) are employed to separate ions based on their mass-to-charge ratios (m/z). TOF measures the time taken for ions to reach the detector, whereas Quadrupole and Ion Trap analyzers filter ions based on m/z ratios via electric or magnetic fields. Orbitrap, another prevalent analyzer, traps ions and measures their frequencies to deduce their m/z ratios.

Following mass analysis, fragmentation, often through Collision-Induced Dissociation (CID)(Shukla & Futrell, 2000) or Higher-Energy Collisional Dissociation (HCD)(Schuhmann et al., 2011), is employed in MS/MS. MS/MS or MS2 is an analysis employing two (or more) mass analyzers connected by a collision cell in a single instrument.

During the elution process by applying LC technique, peptides that possess comparable physicochemical characteristics are separated from the column. Subsequently, the mass spectrometer performs survey scans to detect and pick up the peptide ions that exhibit the highest abundance to tandem mass-spectrometry (Edwards, 2011). For the first stage, an electrospray ionizer generates peptide ions, which are then propelled towards the first mass analyzer (MS1). This analyzer functions as a mass filter enabling the selection of a singular peptide ion (precursor ion) by its mass-to-charge ratio (m/z or m/Q). CID involves collisions between precursor ions and neutral gas molecules, resulting in peptide fragmentation and the generation of sequence-informative ions. HCD, frequently utilized in Orbitrap instruments, employs higher collision energies, leading to informative fragmentation patterns for robust identification in MS2 spectra. This analysis provides information about the m/z values of precursor as well as product ions, thus unlocking a key to protein identification (Eriksson & Fenyö, 2011).

These steps form a cohesive process in MS-based peptide identification, enabling the elucidation of peptide sequences and modifications critical for comprehensive bottom-up proteomic analysis. This intricate workflow of MS analysis holds tremendous potential in understanding the complex proteomic landscape, unraveling the intricacies of protein structure, function, and interactions within biological systems.

Data Acquisition Using Mass Spectrometry

In the landscape of proteomic investigations, data acquisition strategies can be broadly classified as discovery or targeted approaches.

Discovery proteomics involves comprehensive analysis of a sample, aiming to identify/quantify as many peptides as possible per measurement time. This approach utilizes either a data-dependent acquisition (DDA), a data acquisition strategy which select few top intensity ions in each MS1 scan cycle for fragmentation and MS2 scan, or a data-independent acquisition (DIA), where all the ions selected in MS1 scan are fragmentated and scanned for their MS2 irrespective of their intensities (Ravuri et al., 2022). Despite their advantages, DDA have some limitations in terms of reproducible identification and quantification of low-abundance peptides such as kinases, transcription factors, or specific post-translationally modified peptides. DDA are biased towards the detection of more abundant peptides. This constrains the widespread application of discovery methods in high-throughput basic research or clinical environments (Ebhardt et al., 2015; Faria et al., 2017; Picotti et al., 2013; Shi et al., 2016). With the advent of DIA strategies in proteomics (like SWATH-MS, dia-PASEF, etc.), it is now possible to identify and quantify relatively low abundant proteins and do a more reliable and reproducible quantification. Acquired data is processed and searched against databases for peptide-based protein identification and quantitative information about peptide abundance can be derived optionally. As a result, this comprehensive procedure enables the identification, characterization, and quantification of proteins within complex samples (Karpievitch et al., 2010).

Conversely, targeted proteomics homes in on specific proteins or peptides of interest, employing multiple reaction monitoring (MRM) (Lange et al., 2008) or parallel reaction monitoring (PRM) (Peterson et al., 2012) to precisely quantify predefined sets of target peptides. MRM, also known as selected reaction monitoring (SRM), offers high specificity and sensitivity, making it highly suitable for biomarker validation and quantitative analysis. Quantitation of proteins using MRM is achieved through selection of peptides that can acts as a surrogate of target protein. Unique precursor and fragments ion pairs (transitions), generally 4-5 per peptide are selected and analyzed in mass spectrometer which offers highly specific and selective quantitation.

PRM extends this concept by enabling high-resolution and accurate quantification of specific peptides, with improved selectivity and precision. In a typical SRM or PRM run, the mass spectrometer consistently collects spectra at the expected mass-to-charge (m/z) ratio and chromatographic retention time (RT) of the target peptides. The execution of MRM or SRM takes place using triple quadrupole mass spectrometers, wherein the first quadrupole is used for the selection of precursor ions, while the second quadrupole also referred to as "collision cell" is utilized for their fragmentation. Subsequently, the third quadrupole is used to selectively isolate fragment ions that are particular to the target protein, which are then subjected to detection. In practice, many fragment ions can be consecutively observed. Triple quadrupole mass spectrometers provide a high level of durability and are comparatively cost-effective, hence facilitating the integration of SRM/MRM assays within a clinical environment (Whiteaker et al., 2021; Zhang et al., 2018). In contrast to SRM, investigations using PRM are conducted using advanced analytical instruments capable of capturing whole fragment spectra, such as Quadrupole-Orbitrap mass spectrometers and QqTOF systems. In PRM, the quantification of all fragment ions derived from a certain precursor ion is performed simultaneously. In SRM or PRM configurations, an inclusion list is provided to the mass spectrometer, specifying the precursor mass-to-charge ratio (m/z) windows and retention time (RT) windows. Additional scan parameters may be included for certain systems, such as scan-specific fragmentation energies or injection timings. For SRM, the inclusion list serves the purpose of specifying the m/z values of the fragment ions that are to be monitored (Rauniyar, 2015). Therefore, SRM and PRM excel in targeted analyses by focusing on specific precursor-product ion pairs, achieving high sensitivity and reproducibility in quantification.

Data Analysis

Protein Identification: Database and Scoring Alrogithms

For identifying proteins in a sample, database searching is the linchpin of this process. The heart of this technique lies in the marriage of experimental data with theoretical knowledge encoded in protein sequence databases. As a starting point, a protein sequence database is compiled, encompassing the known amino acid sequences of proteins from the relevant organism or database references. Enzyme-specific digestion rules are applied to generate a plethora of predicted peptides. These peptides undergo simulated insilico fragmentation to produce theoretical fragment ions, commonly referred to as "b-ions" and "y-ions"(Michalski et al., 2012). These ions are fragments produced through the controlled fragmentation of peptides and carry essential sequence information b ions have N-terminal of the peptide while y ions possess C-terminus of the peptide. The specific mass-to-charge ratio (m/z) values of b and y ions are measured accurately by the mass spectrometer, enabling their use as signatures for peptide identification. By comparing the observed m/z values of these ions with those expected based on known amino acid sequences in protein databases, researchers can deduce the peptide sequence. The presence of a complete or near-complete set of matching b and y ions in the MS/MS spectrum enhances the confidence in peptide identification. This approach not only enables the precise identification of peptides but also facilitates the discovery of post-translational modifications and elucidation of complex protein mixtures, thus playing a pivotal role in advancing our understanding of proteomic landscapes (Sobott et al., 2011).

Computational analysis often starts by identifying the peptides present in the obtained MS/MS spectra. Search of MS/MS spectra or data against protein sequence database is the most efficient peptide identification method in high-throughput studies (Nesvizhskii, 2006). There are various algorithms and computational strategies used to assess the alignment between the two sets of ions, SEQUEST (Sadygov, 2015), X!Tandem (Craig & Beavis, 2004), and Mascot (Perkins et al., 1999) have been developed to facilitate the data analysis process and to enhance the accuracy of peptide identification. These algorithms can be broadly categorized into database searching algorithms, scoring functions, and post-processing methods.

SEQUEST: Cross-Correlation and High-Throughput Efficiency

As described above, SEQUEST was one of the pioneering algorithms designed for peptide identification in MS-based proteomics. Its strength lies in its use of the cross-correlation (Xcorr) scoring system, which evaluates how closely an experimental spectrum matches the theoretical spectrum of a peptide. This score is calculated by shifting the experimental spectrum against the theoretical spectrum in small increments and computing the correlation at each step. The maximum correlation score becomes the Xcorr value, which is used to rank peptide candidates (Eng et al., 2015).

SEQUEST then ranks candidate peptide sequences based on these scores, with higher scores indicating stronger evidence for identification. A key aspect of SEQUEST's power is its ability to handle large datasets efficiently, making it particularly well-suited for high-throughput proteomics. The cross-correlation method is highly effective at filtering out noise from spectra, which is common in complex biological samples, and focusing on the relevant signal corresponding to peptide fragments. This feature is crucial in large-scale studies, where thousands of spectra are generated and need to be processed in a relatively short amount of time. The ability to filter noise and reduce the search space to likely peptide candidates allows SEQUEST to achieve a high level of sensitivity and accuracy, even in complex proteomic samples (Diament & Noble, 2011; Martiánez-Bartolomé et al., 2008; Sobott et al., 2011).

In addition, SEQUEST has flexibility in accounting for post-translational modifications (PTMs). PTMs such as phosphorylation, glycosylation, and ubiquitination play crucial roles in regulating protein function, and their identification is essential for understanding cellular processes. SEQUEST allows users to specify variable modifications in the search parameters, enabling the algorithm to consider these modifications when generating theoretical spectra. This flexibility makes SEQUEST a valuable tool for studies focusing on dynamic biological processes where PTMs are prevalent. Additionally, SEQUEST can incorporate information about the precursor ion mass tolerance, which defines the allowable mass difference between the observed precursor ion mass and the theoretical mass of the peptide. This feature is particularly important for MS instruments with lower mass accuracy (Bugyi et al., 2021; Tsai et al., 2011) .

SEQUEST has some limitations, particularly in its reliance on the accuracy of precursor ion mass measurements. If the precursor mass is inaccurately measured, the search results can be affected, leading to lower confidence in peptide identifications. To mitigate this issue, SEQUEST results often undergo post-processing validation, such as false discovery rate (FDR) estimations, which help to control the rate of false positives in large-scale experiments. SEQUEST itself does not provide probabilistic scoring or confidence metrics for peptide-spectrum matches (PSMs) like some newer algorithms (e.g., Mascot), so external tools are frequently used to validate the reliability of the results (Prieto & Vázquez, 2020).

Despite these challenges, SEQUEST remains a widely used tool in proteomics due to its robust performance, flexibility in handling PTMs, and effectiveness in reducing noise and computational time. It is particularly well-suited for large-scale studies requiring a high-throughput solution that balances computational complexity with identification accuracy (Gholamizoj et al., 2022).

Mascot: Probability-Based Scoring and Statistical Robustness

Mascot, developed by Matrix Science, focuses on probability-based scoring. Rather than relying on cross-correlation, Mascot calculates the probability that the match between an observed experimental spectrum and a theoretical spectrum occurs by chance. This probabilistic framework allows Mascot to assign a score referred to as the Mascot ion score—which reflects the likelihood that the observed match is genuine. By using this scoring method, Mascot provides a statistically robust way to evaluate peptide-spectrum matches (PSMs), offering a clearer sense of the confidence level associated with each identification (Bjornson et al., 2007).

At the heart of Mascot's scoring mechanism is the expectation value (E-value), which represents the probability of observing a match as good as or better than the current one by random chance. A lower E-value indicates a more significant match, while higher E-values suggest that the match is likely to be random. This statistical approach is one of the key factors that distinguish Mascot from SEQUEST and X! Tandem. In Mascot, the observed score for a peptide-spectrum match is compared to the scores expected from random matches, and only those PSMs that surpass a certain significance threshold are considered valid. This methodology enhances the accuracy and reliability of peptide identifications, particularly in complex datasets (Chen et al., 2019; Lin et al., 2014).

Besides, versatility is one of the primary advantages of Mascot. The algorithm is designed to work across a broad spectrum of MS platforms and supports a wide range of search parameters. Mascot allows users to specify the enzyme used for protein digestion (such as trypsin), the precursor mass tolerance, and various variable modifications that might be present in the sample, such as phosphorylation or acetylation. This flexibility makes Mascot highly adaptable to different experimental setups, from discovery-based proteomics to more targeted analyses, such as biomarker validation in clinical studies. Additionally, Mascot is compatible with a wide variety of protein sequence databases, enabling its use in studies involving diverse organisms or specific biological contexts (Brosch et al., 2008; Mishto et al., 2022). Furthermore, Mascot lies in its ability to handle large and complex protein sequence databases. As the field of proteomics continues to expand, the availability of genomic and proteomic data from various organisms has grown exponentially. Mascot is well-suited to searching these large databases, allowing researchers to analyze data across a wide range of biological systems. The algorithm's ability to filter and rank peptide candidates based on their E-value ensures that only the most statistically significant matches are reported, minimizing the occurrence of false positives (Mishto et al., 2022).

Mascot's reliance on database quality and search parameters means that it is highly dependent on the accuracy of the input data. Incomplete or poorly curated databases can lead to lower confidence in peptide identifications, as the algorithm may fail to find accurate matches for the observed spectra. Similarly, if the search parameters are not carefully optimized (e.g., mass tolerance or enzyme specificity), Mascot's performance can suffer. This highlights the importance of using well-curated databases and carefully tuning search parameters for each experiment (Bugyi et al., 2021; Keenan et al., 2021). Mascot also requires more computational resources compared to some other algorithms, particularly when handling very large datasets or complex samples. The probabilistic framework that makes Mascot so powerful also makes it slower than SEQUEST or X! Tandem when processing high-throughput data. This trade-off between statistical robustness and processing speed can be a limiting factor in large-scale proteomics studies, where the sheer volume of data can slow down the analysis. However, for studies requiring high confidence in peptide identifications and a robust statistical framework, Mascot remains a preferred tool (Fitzgibbon et al., 2007).

X!Tandem: Simplicity and Speed in Large-Scale Analyses

X!Tandem was developed to provide a computationally efficient and flexible alternative to SEQUEST and Mascot for peptide identification in MS-based proteomics. While SEQUEST and Mascot focus on complex scoring systems—cross-correlation and probability-based models, respectively, X! Tandem takes a more straightforward approach by using peak matching as its primary method for scoring peptide-spectrum matches (PSMs). This simplified approach allows X! Tandem to process large datasets quickly and efficiently, making it a valuable tool for high-throughput proteomics, especially when computational resources are limited (Bjornson et al., 2007; Craig & Beavis, 2004) .

X!Tandem works by generating theoretical spectra for peptides from a protein database and then comparing these spectra to the observed MS/MS data. The scoring is based on the number of matched peaks, their intensities, and the position of the matches within the spectra. This approach prioritizes speed and efficiency over more sophisticated models, allowing X! Tandem to rapidly identify peptides in large-scale datasets. The simplicity of the scoring model also makes X! Tandem particularly well-suited for datasets with a large number of spectra, such as those generated in large-scale proteomic studies or clinical proteomics(Craig & Beavis, 2004; Fenyö & Beavis, 2003).

One of the major strengths of X! Tandem is its computational efficiency. X! Tandem can process thousands of spectra in a fraction of the time required by SEQUEST or Mascot by focusing on peak matching rather than complex statistical models. This makes it an ideal choice for large-scale studies where time and computational power are critical factors. Additionally, X! Tandem is highly customizable and open-source, meaning that researchers can modify the algorithm to suit their specific needs. For example, users can adjust the scoring criteria or incorporate novel peptide fragmentation techniques to enhance the algorithm's performance in specific experimental setups. (Akhtar et al., 2012; Bjornson et al., 2007).

Despite its speed and flexibility, X!Tandem has some limitations, particularly when it comes to handling PTMs and complex modifications. While SEQUEST and Mascot are well-equipped to account for variable modifications, X! Tandem's simpler scoring model makes it less effective at identifying peptides with modifications, especially those that require more advanced scoring algorithms to detect. As a result, X! Tandem may miss important PTM identifications in studies focused on protein modifications, such as phosphorylation, glycosylation, or acetylation. This limitation can be particularly problematic in studies where PTMs play a key role in the biological processes being investigated (Kitamura & Galligan, 2023). Another potential drawback of X! Tandem is its performance in the presence of noisy or low-quality spectra. While the algorithm is highly effective at processing clean data, it may struggle to identify peptides accurately in datasets with a high degree of noise or spectral artifacts. This is because X! Tandem's peak-matching approach does not incorporate sophisticated noise-filtering algorithms, which are present in more complex models like SEQUEST. As a result, researchers may need to apply additional filtering or validation steps when using X! Tandem to ensure that the results are reliable(McDonnell et al., 2022).

Protein Quantification

The goal of quantitative proteomics is to quantify the abundance of proteins present in complex biological samples like cells and tissue extracts or subcellular fractions. Protein quantification in MS-based proteomics involves the determination of peptide abundances from acquired spectral data, aiding in the quantification of protein expression levels. Several quantitative proteomic approaches based on MS have been developed in the past few years. Quantitative proteomics can be broadly categorized into relative and absolute quantification.

Relative quantification is used in order to determine how the proteome expresses itself differently in two or more samples, of which two main approaches are stable isotope labeling and label-free quantitation. With the help of this relative methodology, the expression of proteins in two distinct biological states can be compared simultaneously, for instance, the comparison of yeast growing on two different media (galactose medium vs glucose medium), or expression in normal cells compared to cancer cells (Chan et al., 2015). While absolute quantification gives much more accurate levels of the proteins involved in biological processes by providing the specific protein or peptide concentration. (Kito & Ito, 2008; Lindemann et al., 2017).

These methods often incorporate statistical methods to estimate the uncertainty associated with these ratios. Normalization algorithms are vital to mitigate variations in peptide quantification arising from technical factors through adjustment of the quantification values by accounting for factors like sample loading, instrument response, and ionization efficiency (Dressler et al., 2022; Wang et al., 2021). Additionally, quality control algorithms assess the reliability of quantification results, flagging spectra with low signal-to-noise ratios, potential interference, or inadequate fragmentation quality (Wang et al., 2021).

Stable Isotope Labeling Methods

Generally, stable isotope labeling methods involve the incorporation of signature tags into proteins or peptides. Several approaches have been used to introduce the tag to certain functional groups of polypeptides, through chemical reaction or metabolic pathway. As a representative for chemical reaction labeling, there is isotope-coded affinity tag (ICAT) method in which the proteins from two different samples are labeled separately using two ICAT reagents which are identical but differ in their isotopes. Usually, one sample is tagged with ^{13}C isotope (heavy) and the other is tagged with normal (light) ^{12}C isotope via selective in vitro alkylation of proteins at the amino acid cysteine under denaturing and reducing states (Chan et al., 2015). Then, two samples are mixed together and undergo digestion with protease, such as trypsin, creating a mixture of peptides, however containing both unlabeled and labeled fractions. Therefore, the separation is performed through strong cation-exchange (SCX) chromatography to selectively isolate the ICAT-labeled peptides which cling on the monomeric avidin column through the biotin tag, while the unlabeled peptides pass through and come out of the column (Chan et al., 2015). After chemically removing the biotin tag, the labeled peptides are analyzed through the LC-MS to produce the MS spectra for relative quantification. The comparison of "light" peptides and "heavy" peptides is performed using MS spectra resulting in different m/z values (Shuken, 2023).

Stable isotope labeling by amino acids in cell culture (SILAC) is another approach relying on the metabolic incorporation of a given "light" or "heavy" form of the amino acid into the proteins of the cells. Basically, cells are grown separately in culture media supplemented with light or heavy versions of amino acid (Ankney et al., 2018; Lindemann et al., 2017). The light could be normal form of leucine, ^{12}C-lysine, or ^{12}C-arginine corresponding to deuterated leucine, ^{13}C-lysine, or ^{13}C-arginine as heavy amino acids (Ong et al., 2002; Pan & Aebersold, 2007; Shuken, 2023). After a certain number of cell division and differentiation, the heavy isotope–labeled amino acids replace all the light ones which are present in the proteome of the cells cultured within the "heavy" media. It has been shown that the growth of cells cultured in these labeled isotope-containing media has no significant differences from the cells grown in normal media with the evidence of cell morphology, doubling time, and ability to differentiate (Ong et al., 2002). The cells from two media are then harvested and combined with the equal population when the labeled amino acids are completely incorporated into the proteins (Xiaoshi et al., 2015). The time needed for a complete incorporation depends on the type of cell lines and experimental conditions. After the cell lysis, proteins from both samples are extracted and estimated using appropriate assay such as Bradford or Bicinchoninic acid (BCA). Subsequently, protein mixture is digested with trypsin or LysC which cleaves at the C-terminus of lysine or arginine (Lindemann et al., 2017; Pan & Aebersold, 2007; Xiaoshi et al., 2015). These peptides are finally analyzed in LC-MS for quantification. The peak intensities of the heavy and light isotope-labeled peptides are then compared to determine the difference of peptide abundance in one sample relative to that of the other (Ankney et al., 2018). Computational tools, such as MaxQuant (Lindemann et al., 2017), is employed to calculate the heavy to light ratio, provides evidence of variations in protein abundance between two samples.

Absolute quantification through isotope labeling (AQUA) is a potent method to determine the precise protein or peptide concentration, in which labeled peptides with known-concentration, considered as an internal standard, are analyzed together with the target sample. The synthetic peptides are designed to be identical to the endogenous proteolytic peptides generated after digestion so that they can be eluted together and simultaneously analyzed by MS/MS. One difference between the two is that the synthetic peptides are labeled with heavy isotopes such as ^{13}C or ^{15}N (Lindemann et al., 2017). The peptide of interest in the sample is mixed with the heavy isotope labelled peptides followed by detection and analysis via selected reaction monitoring (SRM) in a tandem mass spectrometer. The quantification of sample peptides by SRM is based on the peak area ratio of labeled to unlabeled peptide using which concentration of the target protein can be calculated (Ankney et al., 2018; Lindemann et al., 2017). Data analysis can be performed using Skyline, which is a freely available, open-source software tool for quantitative MS, notably SRM-MS analysis (K. J. Adams et al., 2020).

Label-Free Methods

Label-free methods have been progressed as a rapid and affordable alternative to other quantitative proteomic methodologies. These approaches are effective for large-sample analysis in clinical screening or biomarker discovery studies. Although these are effective at detecting large changes in protein expression but lack accuracy in detecting minor changes. Label-free samples are collected, processed, and analyzed by LC-MS or LC-MS/MS individually, which is in contrast to the isotopic labeling quantification methods. Because the procedure is handled separately for two samples, thus, the performance needs to be controlled gingerly in order to restrict the experimental variations. There are two main label-free approaches, peptide peak intensity or spectral counting, both of which rely on proteolytic digestion of a sample prior to mass spectrometric analysis (Gale et al., 2014). Equal amounts of two or more samples undergo proteolytic cleavage, separated through liquid chromatography, and then analyzed via the mass spectrometer. For relative quantification by ion peak intensity, it relies on the peptide ions separated and detected by LC-MS. The MS spectrum provides the peak intensities corresponding to all detected precursor ions at certain m/z values and a particular retention time. Same ions originated from different samples are compared via their peak signal intensities observed from their individual MS spectrum (Gale et al., 2014). There is a correlation between signal intensity and peptide abundance, thus, leading to the estimation of relative peptide concentration between samples (Ankney et al., 2018). Basically, the higher concentration of proteins presents in a sample, the greater number of peptides would be generated upon digestion, leading to the higher intensity of the peptide ions detected through the mass spectrometer. This method also needs some computer algorithms for ion alignment, comparative analysis and even accounting for bias in peak intensity (intensity normalization) (Ankney et al., 2018). Another label-free approach is spectral counting based on MS/MS method. The number of MS/MS spectra detected from peptides originating from the same protein but from different samples are compared for relative quantification. The correlation between spectral counts and protein abundance is based on the fact that the higher abundance of protein would result in the more number of peptides generated upon digestion, these peptides then undergoes further fragmentation to produce more spectra than the sample with lower abundance of protein (Ankney et al., 2018). It does not require specialized computer tools or algorithms for analysis like the peak intensity method, however, to lower the variance of the data, intensity normalization is still needed.

Applications of Mass Spectrometry-Based Proteomics

The identification, characterization, and quantification of proteins, which are crucial components of life-sustaining activities, are now possible through the application of MS that has been proved to be a highly effective method. Proteome-level identification and characterization contribute to deeper comprehension of the core concepts of protein biology that can lead to the solution of several unsolved biomedical problems. The incorporation of MS instruments and chromatographic separation techniques, such as HPLC, in combination with data analysis programs displays a great potential for advancing our knowledge in proteomics or more specifically in life science. This section highlights the basic applications of MS in protein quantification and analysis of post-translational modifications along with the applications in medical and research areas, notably biomedical research, drug discovery, and even disease diagnosis.

Discovery of Disease- Associated Biomarkers

MS is applied for proteomic analysis of wide range of samples, such as animal tissues or cultured cells, to understand biological processes (Darie-Ion et al.,

2022). There are two different methods for analyzing tissue biopsy samples, in tissue and off tissues. With in-tissue proteomics, an approach called mass spectrometry imaging (MSI) is employed to not only detect the target proteins but also map the spatial distribution of proteins in tissue sections, such as fresh frozen (FF) or formalin-fixed paraffin-embedded (FFPE) tissues. After euthanasia of a model animal, or surgical procedures, the tissue biopsy samples are processed immediately through dissection for cryosection preparation (FF or FFPE). Samples are then micro-spotted for in-tissue trypsin digestion and other preparation steps in order to be analyzed under MS/MS for research or diagnostic purposes. On the other hand, off-tissue method has similar sample preparation strategy to the common proteomics workflow. The tissue sample is homogenized, lysed, and digested before undergoing LC-MS/MS for detection (Darie-Ion et al., 2022). These methods were practically applied, for instance, utilized for thorough understanding of hepatitis B virus (HBV)-associated liver fibrosis by analyzing the liver tissue of rat model infected with HBV (Kan et al., 2017), or examining the Sertoli cell-only syndrome with testis tissue from patients that have no sperm in the ejaculate (Jingping et al., 2012).

In addition, biofluids, such as plasma, urine, or saliva, can also be used for MS/MS analysis, and it contribute a pivotal role in the biomarker discovery for disease diagnosis and prognosis. For instance, blood proteome from cancer patients, such as ovarian or breast cancer, and from healthy subjects were compared to observe differentially expressed proteins that might serve as disease biomarkers; or, saliva was collected for quick screening to detect oral pathologies and distinguish it from other diseases (Darie-Ion et al., 2022). Moreover, sputum was seized to study the pathogenic mechanism of tuberculosis and seek out biomarkers for diagnosis of pulmonary tuberculosis infection (Fu et al., 2012). Other researchers utilized human follicular fluid proteome to be knowledgeable about the follicular formation and oocyte maturation, additionally identifying specific proteins for diagnosing female infertility and the success of in vitro fertilization (Shen et al., 2017).

Development of Drugs

Nearly all biological activities in the human body are regulated by proteins. This indicates that a thorough examination of the proteins offers an opportunity to learn more about their function, action, and interaction with other biomolecules. Accordingly, aberrant alterations in protein expression during disease conditions could display a preliminary prediction of abnormal biomolecular pathways and prospective medicine targets. Till now, most of the drug-based studies considered proteins as target attachment for inhibition, degradation, or triggering (Meissner et al., 2022). With the assistance of MS-based proteomics, many biological mechanisms have been comprehended that aid the determination of potential drug targets, support in developing pertinent medicines, and discovery of new drugs. Specifically, MS helps in detection of covalent binding between drug molecules and target proteins (Loo et al., 1999), with the process similar to PTM analysis. Furthermore, characterization of proteins through MS may highly contribute towards discovering new biological drugs, for instance, antibodies, conjugates of antibodies, drug, and synthetic proteins. These prospective drug molecules are notably more complex than the other chemical drugs, hence, more advanced mass spectrometric methods are needed to efficiently characterize them. Therefore, the emergence of MS-based proteomics significantly with the advancement of MRM has impulsed the discovery process of novel biomarkers which may take an important role in biological systems as well as become effective in disease diagnosis and treatment.

Understanding of Molecular Pathways

Characterization of Post-Translational Modifications

Post-translational modifications (PTM) of proteins have a significant role in the regulation of cellular processes, such as protein folding, localization, or function assignment, which contribute to all the major events like growth, division, differentiation, and metabolism of the cells (Nanni et al., 2022; Carol E. Parker et al., 2010). Specifically, PTMs alter the polypeptide chains after translation via either supplementing or removing certain chemical moieties, cleaving several terminals, or linking two or more protein domains by covalent bond (Larsen et al., 2006). Nowadays, MS has been acknowledged as a great assistance for PTM investigations. The newly formed covalent bonds as the outcome of PTM, can alter the molecular mass of peptides or amino acids specifically, leading to mass enhancement or curtailment, which can be detected via MS (Larsen et al., 2006). Although the fact that a variety of PTMs has been reported in molecular science, only a small number of them have gained extensive attraction. Some of the most notable ones are protein phosphorylation, glycosylation, acetylation, and ubiquitination (Nanni et al., 2022). Sample preparation for PTM analysis is not much different from the general procedure, however, in case of a lower number of peptides modified by PTM, a specific enrichment process should be performed to enable comprehensive analysis. The complexity of intact protein also needs to be fractionated into smaller peptides before going through the mass spectrometer, particularly when tandem mass spectrometer is used, for obtaining MS spectra. The presence of PTM can be detected by the mass shift due to the modified residues.

Conceptually, the peptide with amino acid residues modified with PTM will move further on the m/z values than the non-modified peptides. As in Figure 5, the precursor peptides 1 and 2 represent the non-PTM peptide and PTM modified peptide, respectively. The MS2 spectra (Figure 5) display that PTM leads to an increase in mass and also in m/z values of peptides. (Larsen et al., 2006; Nanni et al., 2022)

Different types of PTMs correspond to adding/removing different chemical moieties by specific enzymes. With phosphorylation, the addition of a phosphate group plays an important role in controlling protein activity and signal transmission (Nanni et al., 2022). As a result of this addition, phosphopeptides can gain 80 Dalton (Da) per modified residue compared to the non-phosphorylated one (Larsen et al., 2006; Carol E. Parker et al., 2010). For glycosylation, residues are attached to glycans with the common type O-linked-glycosylation, which regulates protein structure (folding), function and stability (Nanni et al., 2022). The mass change due to glycosylation is more than 203 Da per modification between modified and unmodified peptides (Larsen et al., 2006). Protein ubiquitination, which is involved in protein degradation or breakdown, result in the mass shift of around 114.04 Da.

In practice, the comparison is performed between the mass of detected peptide in sample and the mass calculated from summing all the amino acid residues of the peptides, then the change in mass would reflect type of modification present (Larsen et al., 2006). This can be achieved through the application of tools such as SEQUEST (Larsen et al., 2006) for instance, which analyzes the unknown MS spectra and matches them to the theoretical spectra of known peptide in the sequence database in order to identify the protein. Therefore, by analyzing MS1 associated with MS2 spectra, it may also detect the amino acids being modified along with the site of modifications in the sequence.

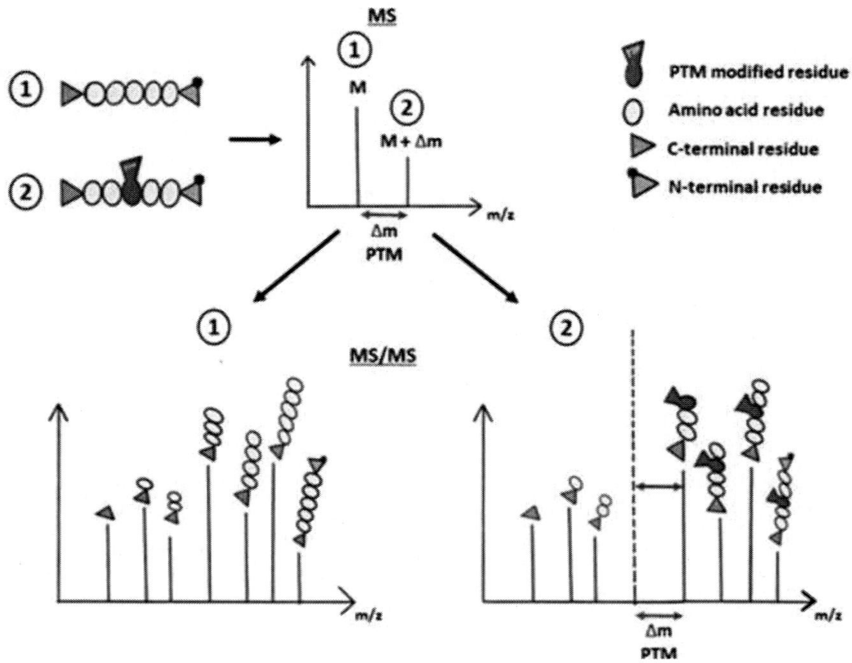

Figure 5. The fundamental workflow for mass spectrometry-based analysis of post-translational modifications (PTMs). Figure is generated using Biorender free version 2024.

Protein-Protein Interaction Studies

Protein-protein interaction studies using MS have become a powerful and widely used approach in the field of proteomics. MS allows for the identification and quantification of proteins within complexes, providing valuable information about protein-protein interactions. The purification of bait-prey complex coupled with MS analysis facilitates the identification of interacting proteins. The method demands the isolation of proteins of interest (bait), which interact with other proteins (prey), and the identification of bait–prey complex components. Therein, affinity purification combined with MS is a well-known technique which is commonly used to identify the interaction.

With affinity purification, the workflow includes affinity tagging, cell lysis, protein complex isolation, and purification followed by MS identification. For this procedure, after cell lysis, protein complex can be isolated and purified by immunoprecipitation technique using antibody that binds to the epitope of target protein (Dunham et al., 2012). This method has the advantage of not manipulating the gene of the target protein, which may be the key in preserving several particular bindings between proteins (Chen & Chen, 2021). However, the collaboration between protein of interest and antibody is not always available or specific and even the high cost for antibody could be its limitations (Chen & Chen, 2021). Therefore, the alternative is epitope tagging that can produce particular epitope tag on target protein which is specific to a certain matrix or resin containing immobilized ligands or antibody for purification. Some of the commonly used tags are GFP (green fluorescent protein), FLAG-tag, Strep-tag (Richards et al., 2021). Gene editing techniques are required for the purpose of fusing DNA sequences to an Open Reading Frame so that the fused protein with the tag can be expressed on the protein of interest (Dunham et al., 2012). After cell lysis, the lysate is incubated with the specific antibody or epitope recognizing matrix. Then, non-specifically bound proteins are washed away keeping only the target protein complexes binding to the matrix. The bait proteins with their interacting prey are then eluted by disrupting the interaction between epitope and matrix (Dunham et al., 2012). The eluted proteins are subjected to MS analysis in order to identify proteins present in the sample.

MS has been widely applied to unravel complex protein interaction networks and understand the functional roles of proteins within the cells. Besides, it can also contribute to the studies of human diseases by being the key approach to provide insights about the changes of protein interaction upon mutations or pathogen infections. This can advance our comprehension of disease etiology, progression, and pathogenesis which helps determine better treatment strategies (Richards et al., 2021).

Recent Advancements in Mass Spectrometry-Based Proteomics

In recent years, MS-based proteomics has seen remarkable advancements in both data acquisition strategies and analysis software, enhancing the depth and precision of proteome analysis. Traditional DDA methods have been complemented and, in some cases, surpassed by DIA techniques, providing broader and more reliable coverage of the proteome. Alongside these developments, the introduction of advanced software tools such as MaxQuant, MSFragger, and Skyline has revolutionized the way researchers process and interpret MS data. These tools not only optimize the performance of DDA and DIA data but also provide enhanced sensitivity, speed, and flexibility for complex experiments. Furthermore, recent breakthroughs in the field—such as single-cell proteomics and new DIA acquisition strategies—are pushing the boundaries of what is possible in proteomic research.

MaxQuant: Comprehensive Analysis Platform for DDA and DIA

MaxQuant is one of the most widely utilized software platforms for MS data analysis. Initially designed for DDA workflows, it has since expanded to support DIA data, offering flexibility for various experimental setups. MaxQuant integrates a comprehensive suite of tools for peptide and protein identification, quantification, and statistical validation, making it ideal for large-scale proteomics projects (Prianichnikov et al., 2020; Sinitcyn et al., 2021).

MaxQuant can handle label-free quantification (LFQ), allowing researchers to quantify protein abundance across multiple samples without the need for isotope labeling. This makes it particularly valuable in discovery-based proteomics, where large numbers of samples are analyzed to identify differential protein expression patterns. Additionally, MaxQuant supports various labeling techniques, including Tandem Mass Tags (TMT) and Isobaric Tags for Relative and Absolute Quantitation (iTRAQ), enabling researchers to perform multiplexed quantification across multiple conditions (Ammar et al., 2023; Bai et al., 2023; Cox et al., 2014).

MaxQuant's Andromeda search engine is another major feature, providing high sensitivity in matching experimental spectra to theoretical spectra from protein databases. It accounts for variable modifications, multiple enzymes, and different cleavage rules, making it versatile for a wide range of experimental designs. For DIA data, MaxQuant integrates with tools like DIA-NN (Data-Independent Acquisition by Neural Networks), which enhances peptide identification and quantification by employing machine learning algorithms. This allows researchers to achieve greater depth of coverage and more accurate quantification, even in highly complex datasets (Cox et al., 2011; Sinitcyn et al., 2021).

Despite its strengths, MaxQuant requires significant computational resources, particularly when dealing with large-scale proteomics data. However, the combination of its powerful search engine, flexibility in handling various experimental designs, and ability to integrate DIA workflows makes it a highly valuable tool in modern proteomics research (Sinitcyn et al., 2021).

MSFragger: Speed and Flexibility for High-Throughput Proteomics

MSFragger is a fast and highly versatile search engine that has gained significant popularity in proteomics, particularly for its ability to rapidly analyze large datasets. Unlike traditional search engines that rely on peptide-spectrum matching (PSM) (Polasky et al., 2023), MSFragger uses an open search algorithm, which significantly speeds up the process of identifying peptides. This approach allows MSFragger to explore a much broader search space by considering all possible precursor masses, rather than limiting the search to a predefined mass window (Yang et al., 2023; Yu et al., 2020).

MSFragger is particularly well-suited for analyzing DIA data, where the sheer volume of spectra can be overwhelming for slower search engines. It can process thousands of spectra in a short period, making it ideal for high-throughput studies where time is a critical factor. Additionally, MSFragger supports a wide range of PTMs, including glycosylation, phosphorylation, and ubiquitination, allowing researchers to study dynamic cellular processes with greater precision (Yu et al., 2023).

MSFragger has the ability to process high-resolution data produced by contemporary mass spectrometers. By utilizing the precise mass accuracy of these instruments, MSFragger enables more reliable peptide identifications, even in complex samples that exhibit a broad dynamic range of protein concentrations. This makes it especially beneficial for studies aimed at detecting low-abundance proteins or peptides with PTMs that may be overlooked by traditional search engines (Yu et al., 2023). Moreover, MSFragger excels in its adaptability to various proteomics workflows, supporting both DDA and DIA approaches. It can be seamlessly integrated with tools like FragPipe and Philosopher, which provide efficient pipelines for protein identification, quantification, and statistical analysis. This combination of flexibility, speed, and the ability to manage complex modifications positions MSFragger as a critical asset in modern proteomics research (Sun et al., 2024; Yu et al., 2023).

Skyline: Precision in Targeted Proteomics

Skyline is another widely used tool in the field of MS, particularly for targeted proteomics and quantitative analysis. Initially developed for SRM and PRM workflows, Skyline has since expanded to support DIA, making it a versatile tool for both discovery-based and targeted proteomics (Kendra J. Adams et al., 2020; Kulyyassov, 2021).

Skyline provides precise quantification of specific peptides across multiple samples. It allows researchers to build targeted assays for proteins of interest, using either experimental data or theoretical predictions (Pino et al., 2017). This makes it particularly valuable for biomarker validation studies, where specific proteins are quantified across different experimental conditions or clinical cohorts. Skyline's interface allows users to visualize and validate chromatographic peaks, ensuring high confidence in peptide quantification (Kirkwood et al., 2022).

In DIA workflows, Skyline integrates with tools like OpenSWATH and DIA-NN, providing detailed quantification and analysis of DIA data (Lou et al., 2023; Zhang et al., 2023). This enables researchers to achieve a high degree of sensitivity and specificity in their analyses, even when dealing with complex samples. Additionally, Skyline supports the use of external spectral libraries, allowing researchers to compare their data against well-curated reference datasets, further enhancing the accuracy of peptide identification and quantification (Deutsch et al., 2018). Skyline also supports for multiple MS instruments, making it compatible with a wide range of MS platforms. This versatility, combined with its precision in targeted assays, makes Skyline a powerful tool for both academic and clinical research applications (Marsh et al., 2021).

Recent Developments in Data-Independent Acquisition (DIA)

The growing popularity of DIA methods has led to significant advancements in proteomics, providing deeper coverage of the proteome and more consistent quantification across samples. Unlike traditional DDA methods, where only the most abundant ions are selected for fragmentation, DIA fragments all ions within a specified mass-to-charge (m/z) range, ensuring that even low-abundance peptides are analyzed in every scan. This comprehensive approach has made DIA particularly valuable in large-scale studies, where reproducibility and coverage are critical factors (Lou & Shui, 2024).

SWATH-MS (Sequential Window Acquisition of All Theoretical Mass Spectra) is the used widely in DIA. SWATH-MS divides the entire m/z range into smaller windows and fragments all precursor ions within each window. This approach ensures that no ions are excluded from analysis, providing a more complete and reproducible dataset compared to DDA. SWATH-MS is especially valuable in studies focused on quantitative proteomics, as it allows researchers to compare protein expression levels across multiple conditions with high accuracy (González-Blanco et al., 2024; Sarkisian & Rodda, 2024).

While SWATH-MS is a widely used DIA strategy, it is not the only option available. Methods like dia-PASEF (Parallel Accumulation Serial Fragmentation combined with Data-Independent Acquisition) offer additional benefits in terms of sensitivity and speed. By integrating ion mobility separation with DIA, dia-PASEF enhances ion utilization efficiency, allowing for more sensitive detection of low-abundance peptides. This approach has been shown to provide deeper proteome coverage compared to traditional SWATH-MS, making it particularly suited for studies requiring high sensitivity and precision (Meier et al., 2021; Skowronek et al., 2023).

Moreover, DIA can integrate machine learning tools, such as DIA-NN and ALFQ, which enhance peptide identification and quantification by using neural networks and other computational techniques. These tools are capable of processing large and complex datasets with greater accuracy, reducing the number of false positives and improving the overall quality of the analysis. Machine learning algorithms are particularly valuable in DIA workflows, where the sheer volume of data can be overwhelming for traditional search engines (Demichev et al., 2019; Shajari et al., 2024).

Single-Cell Proteomics: A New Frontier

Single-cell proteomics represents a revolutionary advancement in the field of proteomics, enabling researchers to explore the protein composition of individual cells rather than averaging the signals from millions of cells, as is typical in traditional bulk proteomics. This groundbreaking approach is essential for understanding cellular heterogeneity, which has profound implications in various biological processes, including development, disease progression, and therapeutic response. The ability to analyze proteins at the single-cell level provides unprecedented insights into the functional dynamics of individual cells, allowing researchers to uncover cellular behaviors that are often masked in bulk analyses (Arias-Hidalgo et al., 2022; Nitz et al., 2024).

Traditional proteomics methodologies often require significant amounts of starting material, making it challenging to detect low-abundance proteins or analyze small, heterogeneous populations of cells. Single-cell proteomics overcomes these limitations through the development of advanced technologies that facilitate the extraction, amplification, and quantification of proteins from individual cells (Kim et al., 2023; Yang et al., 2019).

TMT-based labeling (Tandem Mass Tags) is a notable technique employed in single-cell proteomics, which allows for the simultaneous quantification of proteins from multiple single cells in a single MS run. By tagging proteins with isobaric labels, researchers can analyze multiple samples simultaneously, improving throughput and sensitivity. This method enables the quantification of protein expression levels and post-translational modifications with high precision, providing valuable insights into the functional states of individual cells (Johnson et al., 2023; Zecha et al., 2019).

In addition, nanodroplet sample preparation techniques have emerged as crucial tools. These methods allow for the efficient extraction and processing of proteins from single cells, minimizing sample loss and contamination. By isolating single cells and encapsulating them in nanodroplets, researchers can carry out high-throughput proteomic analyses with remarkable sensitivity. This is particularly beneficial for studying rare cell populations, such as circulating tumor cells or immune cells that play critical roles in disease but are often overlooked in bulk analyses (Nalla et al., 2024; Petrosius & Schoof, 2023).

Single-cell proteomics has already been applied to a wide range of biological systems, from cancer research to developmental biology. The ability to analyze protein expression at the single-cell level is expected to revolutionize the field of proteomics, providing unprecedented insights into cellular heterogeneity and the molecular mechanisms underlying health and disease.

Acknowledgement

This work is funded by Vietnam National University Ho Chi Minh City (VNU-HCM) under grant number B2024-28-05.

References

Adams, K. J., Pratt, B., Bose, N., Dubois, L. G., St John-Williams, L., Perrott, K. M., ... Thompson, J. W. (2020). Skyline for Small Molecules: A Unifying Software Package for Quantitative Metabolomics. *J Proteome Res*, *19*(4), 1447-1458. https://doi.org/10.1021/acs.jproteome.9b00640

Adams, K. J., Pratt, B., Bose, N., Dubois, L. G., St. John-Williams, L., Perrott, K. M., ... Thompson, J. W. (2020). Skyline for Small Molecules: A Unifying Software Package for Quantitative Metabolomics. *Journal of Proteome Research*, *19*(4), 1447-1458. https://doi.org/10.1021/acs.jproteome.9b00640

Akhtar, M. N., Southey, B. R., Andrén, P. E., Sweedler, J. V., & Rodriguez-Zas, S. L. (2012). Evaluation of Database Search Programs for Accurate Detection of Neuropeptides in Tandem Mass Spectrometry Experiments. *Journal of Proteome Research*, *11*(12), 6044-6055. https://doi.org/10.1021/pr3007123

Ali-Khan, N., Zuo, X., & Speicher, D. W. (2003). Overview of proteome analysis. *Curr Protoc Protein Sci*, *Chapter 22*, 22.21.21-22.21.19. https://doi.org/10.1002/0471140864.ps2201s30

Ammar, C., Schessner, J. P., Willems, S., Michaelis, A. C., & Mann, M. (2023). Accurate Label-Free Quantification by directLFQ to Compare Unlimited Numbers of Proteomes. *Molecular & Cellular Proteomics*, *22*(7). https://doi.org/10.1016/j.mcpro.2023.100581

Ankney, J. A., Muneer, A., & Chen, X. (2018). Relative and Absolute Quantitation in Mass Spectrometry–Based Proteomics. *Annual Review of Analytical Chemistry*, *11*(1), 49-77. https://doi.org/10.1146/annurev-anchem-061516-045357

Arias-Hidalgo, C., Juanes-Velasco, P., Landeira-Viñuela, A., García-Vaquero, M. L., Montalvillo, E., Góngora, R., ... Fuentes, M. (2022). Single-Cell Proteomics: The Critical Role of Nanotechnology. *International Journal of Molecular Sciences*, *23*(12). https://doi.org/10.3390/ijms23126707

Badawy, M. E. I., El-Nouby, M. A. M., Kimani, P. K., Lim, L. W., & Rabea, E. I. (2022). A review of the modern principles and applications of solid-phase extraction techniques in chromatographic analysis. *Analytical Sciences*, *38*(12), 1457-1487. https://doi.org/10.1007/s44211-022-00190-8

Bai, M., Deng, J., Dai, C., Pfeuffer, J., Sachsenberg, T., & Perez-Riverol, Y. (2023). LFQ-Based Peptide and Protein Intensity Differential Expression Analysis. *Journal of Proteome Research*, *22*(6), 2114-2123. https://doi.org/10.1021/acs.jproteome.2c00812

Beeton-Kempen, N. (2010). *Proteomics: Principles, Techniques and Applications.*

Bjornson, R. D., Carriero, N. J., Colangelo, C., Shifman, M., Cheung, K.-H., Miller, P. L., & Williams, K. (2007). X!!Tandem, an Improved Method for Running X!Tandem in Parallel on Collections of Commodity Computers. *Journal of Proteome Research*, *7*(1), 293-299. https://doi.org/10.1021/pr0701198

Brosch, M., Swamy, S., Hubbard, T., & Choudhary, J. (2008). Comparison of Mascot and X!Tandem Performance for Low and High Accuracy Mass Spectrometry and the Development of an Adjusted Mascot Threshold. *Molecular & Cellular Proteomics*, *7*(5), 962-970. https://doi.org/10.1074/mcp.M700293-MCP200

Bugyi, F., Szabó, D., Szabó, G., Révész, Á., Pape, V. F. S., Soltész-Katona, E., ... Drahos, L. (2021). Influence of Post-Translational Modifications on Protein Identification in Database Searches. *ACS Omega*, *6*(11), 7469-7477. https://doi.org/10.1021/acsomega.0c05997

Chan, J. C. Y., Zhou, L., & Chan, E. C. Y. (2015). The Isotope-Coded Affinity Tag Method for Quantitative Protein Profile Comparison and Relative Quantitation of Cysteine

Redox Modifications. *Curr Protoc Protein Sci*, *82*, 23.22.21-23.22.19. https://doi.org/10.1002/0471140864.ps2302s82

Chen, Y.-L., Chang, W.-H., Lee, C.-Y., & Chen, Y.-R. (2019). An improved scoring method for the identification of endogenous peptides based on the Mascot MS/MS ion search. *The Analyst*, *144*(9), 3045-3055. https://doi.org/10.1039/c8an02141d

Chen, Z., & Chen, J. (2021). Mass spectrometry-based protein–protein interaction techniques and their applications in studies of DNA damage repair. *J Zhejiang Univ Sci B*, *22*(1), 1-20. https://doi.org/10.1631/jzus.B2000356

Chi, A., Bai, D. L., Geer, L. Y., Shabanowitz, J., & Hunt, D. F. (2007). Analysis of intact proteins on a chromatographic time scale by electron transfer dissociation tandem mass spectrometry. *International Journal of Mass Spectrometry*, *259*(1-3), 197-203. https://doi.org/10.1016/j.ijms.2006.09.030

Cox, J., Hein, M. Y., Luber, C. A., Paron, I., Nagaraj, N., & Mann, M. (2014). Accurate Proteome-wide Label-free Quantification by Delayed Normalization and Maximal Peptide Ratio Extraction, Termed MaxLFQ. *Molecular & Cellular Proteomics*, *13*(9), 2513-2526. https://doi.org/10.1074/mcp.M113.031591

Cox, J. r., Neuhauser, N., Michalski, A., Scheltema, R. A., Olsen, J. V., & Mann, M. (2011). Andromeda: A Peptide Search Engine Integrated into the MaxQuant Environment. *Journal of Proteome Research*, *10*(4), 1794-1805. https://doi.org/10.1021/pr101065j

Craig, R., & Beavis, R. C. (2004). TANDEM: matching proteins with tandem mass spectra. *Bioinformatics*, *20*(9), 1466-1467. https://doi.org/10.1093/bioinformatics/bth092

Crankshaw, M. W., & Grant, G. A. (2001). Modification of Cysteine. *Current Protocols in Protein Science*, *3*(1). https://doi.org/10.1002/0471140864.ps1501s03

D'Cunha, R., Bach, T., Young, B. A., Li, P., Nalbant, D., Zhang, J., ... An, G. (2018). Quantification of Cefepime, Meropenem, Piperacillin, and Tazobactam in Human Plasma Using a Sensitive and Robust Liquid Chromatography-Tandem Mass Spectrometry Method, Part 1: Assay Development and Validation. *Antimicrobial Agents and Chemotherapy*, *62*(9). https://doi.org/10.1128/aac.00859-18

Darie-Ion, L., Whitham, D., Jayathirtha, M., Rai, Y., Neagu, A.-N., Darie, C. C., & Petre, B. A. (2022). Applications of MALDI-MS/MS-Based Proteomics in Biomedical Research. *Molecules*, *27*(19), 6196. https://www.mdpi.com/1420-3049/27/19/6196

Dembek, M., & Bocian, S. (2020). Pure water as a mobile phase in liquid chromatography techniques. *TrAC Trends in Analytical Chemistry*, *123*. https://doi.org/10.1016/j.trac.2019.115793

Demichev, V., Messner, C. B., Vernardis, S. I., Lilley, K. S., & Ralser, M. (2019). DIA-NN: neural networks and interference correction enable deep proteome coverage in high throughput. *Nature Methods*, *17*(1), 41-44. https://doi.org/10.1038/s41592-019-0638-x

Deutsch, E. W., Perez-Riverol, Y., Chalkley, R. J., Wilhelm, M., Tate, S., Sachsenberg, T., ... Röst, H. (2018). Expanding the Use of Spectral Libraries in Proteomics. *Journal of Proteome Research*, *17*(12), 4051-4060. https://doi.org/10.1021/acs.jproteome.8b00485

Diament, B. J., & Noble, W. S. (2011). Faster SEQUEST Searching for Peptide Identification from Tandem Mass Spectra. *Journal of Proteome Research*, *10*(9), 3871-3879. https://doi.org/10.1021/pr101196n

Doellinger, J., Schneider, A., Hoeller, M., & Lasch, P. (2020). Sample Preparation by Easy Extraction and Digestion (SPEED) - A Universal, Rapid, and Detergent-free Protocol for Proteomics Based on Acid Extraction. *Molecular & Cellular Proteomics*, *19*(1), 209-222. https://doi.org/10.1074/mcp.TIR119.001616

Dressler, F. F., Brägelmann, J., Reischl, M., & Perner, S. (2022). Normics: Proteomic Normalization by Variance and Data-Inherent Correlation Structure. *Molecular & Cellular Proteomics*, *21*(9). https://doi.org/10.1016/j.mcpro.2022.100269

Dunham, W. H., Mullin, M., & Gingras, A. C. (2012). Affinity-purification coupled to mass spectrometry: basic principles and strategies. *Proteomics*, *12*(10), 1576-1590. https://doi.org/10.1002/pmic.201100523

Ebhardt, H. A., Root, A., Sander, C., & Aebersold, R. (2015). Applications of targeted proteomics in systems biology and translational medicine. *Proteomics*, *15*(18), 3193-3208. https://doi.org/10.1002/pmic.201500004

Edwards, N. J. (2011). Protein Identification from Tandem Mass Spectra by Database Searching. In *Bioinformatics for Comparative Proteomics* (pp. 119-138). https://doi.org/10.1007/978-1-60761-977-2_9

Eng, J. K., Hoopmann, M. R., Jahan, T. A., Egertson, J. D., Noble, W. S., & MacCoss, M. J. (2015). A Deeper Look into Comet—Implementation and Features. *Journal of the American Society for Mass Spectrometry*, *26*(11), 1865-1874. https://doi.org/10.1007/s13361-015-1179-x

Eriksson, J., & Fenyö, D. (2011). Modeling Mass Spectrometry-Based Protein Analysis. In *Bioinformatics for Comparative Proteomics* (pp. 109-117). https://doi.org/10.1007/978-1-60761-977-2_8

Faria, S. S., Morris, C. F. M., Silva, A. R., Fonseca, M. P., Forget, P., Castro, M. S., & Fontes, W. (2017). A Timely Shift from Shotgun to Targeted Proteomics and How It Can Be Groundbreaking for Cancer Research. *Frontiers in Oncology*, *7*. https://doi.org/10.3389/fonc.2017.00013

Fenn, J. B., Mann, M., Meng, C. K., Wong, S. F., & Whitehouse, C. M. (1989). Electrospray Ionization for Mass Spectrometry of Large Biomolecules. *Science*, *246*(4926), 64-71. https://doi.org/10.1126/science.2675315

Fenyö, D., & Beavis, R. C. (2003). A Method for Assessing the Statistical Significance of Mass Spectrometry-Based Protein Identifications Using General Scoring Schemes. *Analytical Chemistry*, *75*(4), 768-774. https://doi.org/10.1021/ac0258709

Fitzgibbon, M., Li, Q., & McIntosh, M. (2007). Modes of Inference for Evaluating the Confidence of Peptide Identifications. *Journal of Proteome Research*, *7*(1), 35-39. https://doi.org/10.1021/pr7007303

Fu, Y. R., Yi, Z. J., Guan, S. Z., Zhang, S. Y., & Li, M. (2012). Proteomic analysis of sputum in patients with active pulmonary tuberculosis. *Clinical Microbiology and Infection*, *18*(12), 1241-1247. https://doi.org/10.1111/j.1469-0691.2012.03824.x

Gale, P. J., Duncan, M. W., & Yergey, A. L. (2014). Quantifying Proteins by Mass Spectrometry. *LCGC North America-09-01-2014*, *32*(9), 726–735. https://www.chromatographyonline.com/view/quantifying-proteins-mass-spectrometry

Gholamizoj, S., Ma, B., & Vitek, O. (2022). SPEQ: quality assessment of peptide tandem mass spectra with deep learning. *Bioinformatics*, *38*(6), 1568-1574. https://doi.org/10.1093/bioinformatics/btab874

Glatter, T., Ludwig, C., Ahrné, E., Aebersold, R., Heck, A. J. R., & Schmidt, A. (2012). Large-Scale Quantitative Assessment of Different In-Solution Protein Digestion Protocols Reveals Superior Cleavage Efficiency of Tandem Lys-C/Trypsin Proteolysis over Trypsin Digestion. *Journal of Proteome Research, 11*(11), 5145-5156. https://doi.org/10.1021/pr300273g

González-Blanco, L., Oliván, M., Diñeiro, Y., Bravo, S. B., Sierra, V., & Gagaoua, M. (2024). Sequential window acquisition of all theoretical mass spectra (SWATH-MS) as an emerging proteomics approach for the discovery of dark-cutting beef biomarkers. *Meat Science, 217*. https://doi.org/10.1016/j.meatsci.2024.109618

Goodman, J. K., Zampronio, C. G., Jones, A. M. E., & Hernandez-Fernaud, J. R. (2018). Updates of the In-Gel Digestion Method for Protein Analysis by Mass Spectrometry. *Proteomics, 18*(23). https://doi.org/10.1002/pmic.201800236

Greco, T. M., Miteva, Y., Conlon, F. L., & Cristea, I. M. (2012). Complementary Proteomic Analysis of Protein Complexes. In *Xenopus Protocols* (pp. 391-407). https://doi.org/10.1007/978-1-61779-992-1_22

Gundry, R. L., White, M. Y., Murray, C. I., Kane, L. A., Fu, Q., Stanley, B. A., & Van Eyk, J. E. (2009). Preparation of Proteins and Peptides for Mass Spectrometry Analysis in a Bottom-Up Proteomics Workflow. In *Current Protocols in Molecular Biology*. https://doi.org/10.1002/0471142727.mb1025s88

Gundry, R. L., White, M. Y., Murray, C. I., Kane, L. A., Fu, Q., Stanley, B. A., & Van Eyk, J. E. (2010). Preparation of Proteins and Peptides for Mass Spectrometry Analysis in a Bottom-Up Proteomics Workflow. *Current Protocols in Molecular Biology, 90*(1). https://doi.org/10.1002/0471142727.mb1025s88

Han, X., Aslanian, A., & Yates, J. R., 3rd. (2008). Mass spectrometry for proteomics. *Curr Opin Chem Biol, 12*(5), 483-490. https://doi.org/10.1016/j.cbpa.2008.07.024

Hill, B. G., Reily, C., Oh, J.-Y., Johnson, M. S., & Landar, A. (2009). Methods for the determination and quantification of the reactive thiol proteome. *Free Radical Biology and Medicine, 47*(6), 675-683. https://doi.org/10.1016/j.freeradbiomed.2009.06.012

Jingping, L., Wenbin, G., Fei, L., Jincan, H., Qingfeng, Y., Xiaoqiang, W., ... Xiangming, M. (2012). HnRNPL as a key factor in spermatogenesis: Lesson from functional proteomic studies of azoospermia patients with sertoli cell only syndrome. *Journal of Proteomics, 75*(10), 2879-2891. https://doi.org/https://doi.org/10.1016/j.jprot.2011.12.040

Johnson, F. D., Hughes, C. S., Liu, A., Lockwood, W. W., & Morin, G. B. (2023). Tandem mass tag-based thermal proteome profiling for the discovery of drug-protein interactions in cancer cells. *STAR Protocols, 4*(1). https://doi.org/10.1016/j.xpro.2022.102012

Kan, F., Ye, L., Yan, T., Cao, J., Zheng, J., & Li, W. (2017). Proteomic and transcriptomic studies of HBV-associated liver fibrosis of an AAV-HBV-infected mouse model. *BMC Genomics, 18*(1), 641. https://doi.org/10.1186/s12864-017-3984-z

Karas, M., & Hillenkamp, F. (2002). Laser desorption ionization of proteins with molecular masses exceeding 10,000 daltons. *Analytical Chemistry, 60*(20), 2299-2301. https://doi.org/10.1021/ac00171a028

Karch, K. R., Sidoli, S., & Garcia, B. A. (2016). Identification and Quantification of Histone PTMs Using High-Resolution Mass Spectrometry. In *Enzymes of Epigenetics, Part B* (pp. 3-29). https://doi.org/10.1016/bs.mie.2015.12.007

Karpievitch, Y. V., Polpitiya, A. D., Anderson, G. A., Smith, R. D., & Dabney, A. R. (2010). Liquid chromatography mass spectrometry-based proteomics: Biological and technological aspects. *The Annals of Applied Statistics*, *4*(4). https://doi.org/10.1214/10-aoas341

Keenan, E. K., Zachman, D. K., & Hirschey, M. D. (2021). Discovering the landscape of protein modifications. *Molecular Cell*, *81*(9), 1868-1878. https://doi.org/10.1016/j.molcel.2021.03.015

Kielkopf, C. L., Bauer, W., & Urbatsch, I. L. (2020). Bradford Assay for Determining Protein Concentration. *Cold Spring Harbor Protocols*, *2020*(4). https://doi.org/10.1101/pdb.prot102269

Kim, S., Kamarulzaman, L., & Taniguchi, Y. (2023). Recent methodological advances towards single-cell proteomics. *Proceedings of the Japan Academy, Series B*, *99*(8), 306-327. https://doi.org/10.2183/pjab.99.021

Kirkwood, K. I., Pratt, B. S., Shulman, N., Tamura, K., MacCoss, M. J., MacLean, B. X., & Baker, E. S. (2022). Utilizing Skyline to analyze lipidomics data containing liquid chromatography, ion mobility spectrometry and mass spectrometry dimensions. *Nature Protocols*, *17*(11), 2415-2430. https://doi.org/10.1038/s41596-022-00714-6

Kitamura, N., & Galligan, J. J. (2023). A global view of the human post-translational modification landscape. *Biochemical Journal*, *480*(16), 1241-1265. https://doi.org/10.1042/bcj20220251

Kito, K., & Ito, T. (2008). Mass spectrometry-based approaches toward absolute quantitative proteomics. *Curr Genomics*, *9*(4), 263-274. https://doi.org/10.2174/138920208784533647

Kulyyassov, A. (2021). Application of Skyline for Analysis of Protein–Protein Interactions In Vivo. *Molecules*, *26*(23). https://doi.org/10.3390/molecules26237170

Lange, V., Picotti, P., Domon, B., & Aebersold, R. (2008). Selected reaction monitoring for quantitative proteomics: a tutorial. *Molecular Systems Biology*, *4*(1). https://doi.org/10.1038/msb.2008.61

Larsen, M. R., Trelle, M. B., Thingholm, T. E., & Jensen, O. N. (2006). Analysis of posttranslational modifications of proteins by tandem mass spectrometry. *BioTechniques*, *40*(6), 790-798. https://doi.org/10.2144/000112201

Lin, M. S., Cherny, J. J., Fournier, C. T., Roth, S. J., Krizanc, D., & Weir, M. P. (2014). Assessment of MS/MS Search Algorithms with Parent-Protein Profiling. *Journal of Proteome Research*, *13*(4), 1823-1832. https://doi.org/10.1021/pr401090d

Lindemann, C., Thomanek, N., Hundt, F., Lerari, T., Meyer, H. E., Wolters, D., & Marcus, K. (2017). Strategies in relative and absolute quantitative mass spectrometry based proteomics. *Biological Chemistry*, *398*(5-6), 687-699. https://doi.org/doi:10.1515/hsz-2017-0104

Loo, J. A., DeJohn, D. E., Du, P., Stevenson, T. I., & Ogorzalek Loo, R. R. (1999). Application of mass spectrometry for target identification and characterization. *Medicinal Research Reviews*, *19*(4), 307-319. https://doi.org/https://doi.org/10.1002/(SICI)1098-1128(199907)19:4<307::AID-MED4>3.0.CO;2-2

Lou, R., Cao, Y., Li, S., Lang, X., Li, Y., Zhang, Y., & Shui, W. (2023). Benchmarking commonly used software suites and analysis workflows for DIA proteomics and phosphoproteomics. *Nature Communications*, *14*(1). https://doi.org/10.1038/s41467-022-35740-1

Lou, R., & Shui, W. (2024). Acquisition and Analysis of DIA-Based Proteomic Data: A Comprehensive Survey in 2023. *Molecular & Cellular Proteomics*, *23*(2). https://doi.org/10.1016/j.mcpro.2024.100712

Makarov, A. (2000). Electrostatic Axially Harmonic Orbital Trapping: A High-Performance Technique of Mass Analysis. *Analytical Chemistry*, *72*(6), 1156-1162. https://doi.org/10.1021/ac991131p

Makarov, A., Denisov, E., Kholomeev, A., Balschun, W., Lange, O., Strupat, K., & Horning, S. (2006). Performance Evaluation of a Hybrid Linear Ion Trap/Orbitrap Mass Spectrometer. *Analytical Chemistry*, *78*(7), 2113-2120. https://doi.org/10.1021/ac0518811

Marsh, A. N., Sharma, V., Mani, S. K., Vitek, O., MacCoss, M. J., & MacLean, B. X. (2021). Skyline Batch: An Intuitive User Interface for Batch Processing with Skyline. *Journal of Proteome Research*, *21*(1), 289-294. https://doi.org/10.1021/acs.jproteome.1c00749

Martiánez-Bartolomé, S., Navarro, P., Martián-Maroto, F., López-Ferrer, D., Ramos-Fernández, A., Villar, M., ... Vázquez, J. (2008). Properties of Average Score Distributions of SEQUEST. *Molecular & Cellular Proteomics*, *7*(6), 1135-1145. https://doi.org/10.1074/mcp.M700239-MCP200

McDonnell, K., Howley, E., & Abram, F. (2022). The impact of noise and missing fragmentation cleavages on de novo peptide identification algorithms. *Computational and Structural Biotechnology Journal*, *20*, 1402-1412. https://doi.org/10.1016/j.csbj.2022.03.008

Medzihradszky, K. F. (2005). In-Solution Digestion of Proteins for Mass Spectrometry. In *Mass Spectrometry: Modified Proteins and Glycoconjugates* (pp. 50-65). https://doi.org/10.1016/s0076-6879(05)05003-2

Meier, F., Park, M. A., & Mann, M. (2021). Trapped Ion Mobility Spectrometry and Parallel Accumulation–Serial Fragmentation in Proteomics. *Molecular & Cellular Proteomics*, *20*. https://doi.org/10.1016/j.mcpro.2021.100138

Meissner, F., Geddes-McAlister, J., Mann, M., & Bantscheff, M. (2022). The emerging role of mass spectrometry-based proteomics in drug discovery. *Nature Reviews Drug Discovery*, *21*(9), 637-654. https://doi.org/10.1038/s41573-022-00409-3

Michalski, A., Neuhauser, N., Cox, J., & Mann, M. (2012). A Systematic Investigation into the Nature of Tryptic HCD Spectra. *Journal of Proteome Research*, *11*(11), 5479-5491. https://doi.org/10.1021/pr3007045

Mishto, M., Horokhovskyi, Y., Cormican, J. A., Yang, X., Lynham, S., Urlaub, H., & Liepe, J. (2022). Database search engines and target database features impinge upon the identification of post-translationally cis-spliced peptides in HLA class I immunopeptidomes. *Proteomics*, *22*(10). https://doi.org/10.1002/pmic.202100226

Nalla, L. V., Kanukolanu, A., Yeduvaka, M., & Gajula, S. N. R. (2024). https://doi.org/10.22541/au.172848272.26849931/v1

Nanni, P., Gehrig, P., & Schlapbach, R. (2022). Mass Spectrometry in Proteomics: Technologies, Methods, and Research Applications for the Life Sciences. *CHIMIA*, *76*(1-2), 73. https://doi.org/10.2533/chimia.2022.73

Nesvizhskii, A. I. (2006). Protein Identification by Tandem Mass Spectrometry and Sequence Database Searching. In *Mass Spectrometry Data Analysis in Proteomics* (pp. 87-120). https://doi.org/10.1385/1-59745-275-0:87

Nika, H., Nieves, E., Hawke, D. H., & Angeletti, R. H. (2013). C-Terminal Protein Characterization by Mass Spectrometry using Combined Micro Scale Liquid and Solid-Phase Derivatization. *Journal of Biomolecular Techniques : JBT*. https://doi.org/10.7171/jbt.13-2401-003

Nitz, A. A., Giraldez Chavez, J. H., Eliason, Z. G., & Payne, S. H. (2024). Are We There Yet? Assessing the Readiness of Single-Cell Proteomics to Answer Biological Hypotheses. *Journal of Proteome Research*. https://doi.org/10.1021/acs.jproteome.4c00091

Ogorzalek Loo, R. R., Cavalcoli, J. D., VanBogelen, R. A., Mitchell, C., Loo, J. A., Moldover, B., & Andrews, P. C. (2001). Virtual 2-D Gel Electrophoresis: Visualization and Analysis of the E. coli Proteome by Mass Spectrometry. *Analytical Chemistry*, *73*(17), 4063-4070. https://doi.org/10.1021/ac0101858

Olsen, J. V., Ong, S.-E., & Mann, M. (2004). Trypsin Cleaves Exclusively C-terminal to Arginine and Lysine Residues. *Molecular & Cellular Proteomics*, *3*(6), 608-614. https://doi.org/10.1074/mcp.T400003-MCP200

Ong, S.-E., Blagoev, B., Kratchmarova, I., Kristensen, D. B., Steen, H., Pandey, A., & Mann, M. (2002). Stable Isotope Labeling by Amino Acids in Cell Culture, SILAC, as a Simple and Accurate Approach to Expression Proteomics *. *Molecular & Cellular Proteomics*, *1*(5), 376-386. https://doi.org/10.1074/mcp.M200025-MCP200

Pan, S., & Aebersold, R. (2007). Quantitative proteomics by stable isotope labeling and mass spectrometry. *Methods Mol Biol*, *367*, 209-218. https://doi.org/10.1385/1-59745-275-0:209

Parker, C. E., Mocanu, V., Mocanu, M., Dicheva, N., & Warren, M. R. (2010). Mass Spectrometry for Post-Translational Modifications. *Neuroproteomics*, *Chapter 6*. https://www.ncbi.nlm.nih.gov/books/NBK56012/?report=classic

Parker, C. E., Warren, M. R., & Mocanu, V. (2010). Mass Spectrometry for Proteomics. In O. Alzate (Ed.), *Neuroproteomics*. https://www.ncbi.nlm.nih.gov/pubmed/21882443

Paulo, J. A. (2016). Sample preparation for proteomic analysis using a GeLC-MS/MS strategy. *Journal of Biological Methods*, *3*(3). https://doi.org/10.14440/jbm.2016.106

Perkins, D. N., Pappin, D. J. C., Creasy, D. M., & Cottrell, J. S. (1999). Probability-based protein identification by searching sequence databases using mass spectrometry data. *Electrophoresis*, *20*(18), 3551-3567. https://doi.org/10.1002/(sici)1522-2683(19991201)20:18<3551::Aid-elps3551>3.0.Co;2-2

Peterson, A. C., Russell, J. D., Bailey, D. J., Westphall, M. S., & Coon, J. J. (2012). Parallel Reaction Monitoring for High Resolution and High Mass Accuracy Quantitative, Targeted Proteomics. *Molecular & Cellular Proteomics*, *11*(11), 1475-1488. https://doi.org/10.1074/mcp.O112.020131

Petrosius, V., & Schoof, E. M. (2023). Recent advances in the field of single-cell proteomics. *Translational Oncology*, *27*. https://doi.org/10.1016/j.tranon.2022.101556

Picotti, P., Bodenmiller, B., & Aebersold, R. (2013). Proteomics meets the scientific method. *Nature Methods*, *10*(1), 24-27. https://doi.org/10.1038/nmeth.2291

Pino, L. K., Searle, B. C., Bollinger, J. G., Nunn, B., MacLean, B., & MacCoss, M. J. (2017). The Skyline ecosystem: Informatics for quantitative mass spectrometry proteomics. *Mass Spectrometry Reviews*, *39*(3), 229-244. https://doi.org/10.1002/mas.21540

Polasky, D. A., Geiszler, D. J., Yu, F., Li, K., Teo, G. C., & Nesvizhskii, A. I. (2023). MSFragger-Labile: A Flexible Method to Improve Labile PTM Analysis in Proteomics. *Molecular & Cellular Proteomics*, *22*(5). https://doi.org/10.1016/j.mcpro.2023.100538

Prianichnikov, N., Koch, H., Koch, S., Lubeck, M., Heilig, R., Brehmer, S., ... Cox, J. (2020). MaxQuant Software for Ion Mobility Enhanced Shotgun Proteomics. *Molecular & Cellular Proteomics*, *19*(6), 1058-1069. https://doi.org/10.1074/mcp.TIR119.001720

Prieto, G., & Vázquez, J. (2020). Protein Probability Model for High-Throughput Protein Identification by Mass Spectrometry-Based Proteomics. *Journal of Proteome Research*, *19*(3), 1285-1297. https://doi.org/10.1021/acs.jproteome.9b00819

Proc, J. L., Kuzyk, M. A., Hardie, D. B., Yang, J., Smith, D. S., Jackson, A. M., ... Borchers, C. H. (2010). A Quantitative Study of the Effects of Chaotropic Agents, Surfactants, and Solvents on the Digestion Efficiency of Human Plasma Proteins by Trypsin. *Journal of Proteome Research*, *9*(10), 5422-5437. https://doi.org/10.1021/pr100656u

Rauniyar, N. (2015). Parallel Reaction Monitoring: A Targeted Experiment Performed Using High Resolution and High Mass Accuracy Mass Spectrometry. *International Journal of Molecular Sciences*, *16*(12), 28566-28581. https://doi.org/10.3390/ijms161226120

Ravuri, H. G., Noor, Z., Mills, P. C., Satake, N., & Sadowski, P. (2022). Data-Independent Acquisition Enables Robust Quantification of 400 Proteins in Non-Depleted Canine Plasma. *Proteomes*, *10*(1). https://doi.org/10.3390/proteomes10010009

Richards, A. L., Eckhardt, M., & Krogan, N. J. (2021). Mass spectrometry-based protein-protein interaction networks for the study of human diseases. *Mol Syst Biol*, *17*(1), e8792. https://doi.org/10.15252/msb.20188792

Rosenfeld, J., Capdevielle, J., Guillemot, J. C., & Ferrara, P. (1992). In-gel digestion of proteins for internal sequence analysis after one- or two-dimensional gel electrophoresis. *Analytical Biochemistry*, *203*(1), 173-179. https://doi.org/10.1016/0003-2697(92)90061-b

Sadygov, R. G. (2015). Using SEQUEST with Theoretically Complete Sequence Databases. *Journal of the American Society for Mass Spectrometry*, *26*(11), 1858-1864. https://doi.org/10.1007/s13361-015-1228-5

Sarkisian, M., & Rodda, L. N. (2024). The use of sequential window acquisition of all theoretical fragment ion spectra (SWATH), a data-independent acquisition high-resolution mass spectrometry approach, in forensic toxicological regimes: A review. *Drug Testing and Analysis*. https://doi.org/10.1002/dta.3700

Schuhmann, K., Herzog, R., Schwudke, D., Metelmann-Strupat, W., Bornstein, S. R., & Shevchenko, A. (2011). Bottom-Up Shotgun Lipidomics by Higher Energy Collisional Dissociation on LTQ Orbitrap Mass Spectrometers. *Analytical Chemistry*, *83*(14), 5480-5487. https://doi.org/10.1021/ac102505f

Sechi, S., & Chait, B. T. (1998). Modification of Cysteine Residues by Alkylation. A Tool in Peptide Mapping and Protein Identification. *Analytical Chemistry*, *70*(24), 5150-5158. https://doi.org/10.1021/ac9806005

Shajari, E., Gagné, D., Malick, M., Roy, P., Noël, J.-F., Gagnon, H., ... Beaulieu, J.-F. (2024). Application of SWATH Mass Spectrometry and Machine Learning in the Diagnosis of Inflammatory Bowel Disease Based on the Stool Proteome. *Biomedicines*, *12*(2). https://doi.org/10.3390/biomedicines12020333

Shen, X., Liu, X., Zhu, P., Zhang, Y., Wang, J., Wang, Y., ... Liu, F. (2017). Proteomic analysis of human follicular fluid associated with successful in vitro fertilization. *Reproductive Biology and Endocrinology*, *15*(1), 58. https://doi.org/10.1186/s12958-017-0277-y

Shevchenko, A., Tomas, H., Havli, J., Olsen, J. V., & Mann, M. (2007). In-gel digestion for mass spectrometric characterization of proteins and proteomes. *Nature Protocols*, *1*(6), 2856-2860. https://doi.org/10.1038/nprot.2006.468

Shi, T., Song, E., Nie, S., Rodland, K. D., Liu, T., Qian, W. J., & Smith, R. D. (2016). Advances in targeted proteomics and applications to biomedical research. *Proteomics*, *16*(15-16), 2160-2182. https://doi.org/10.1002/pmic.201500449

Shuken, S. R. (2023). An Introduction to Mass Spectrometry-Based Proteomics. *Journal of Proteome Research*, *22*(7), 2151-2171. https://doi.org/10.1021/acs.jproteome.2c00838

Shukla, A. K., & Futrell, J. H. (2000). Tandem mass spectrometry: dissociation of ions by collisional activation. *Journal of Mass Spectrometry*, *35*(9), 1069-1090. https://doi.org/10.1002/1096-9888(200009)35:9<1069::Aid-jms54>3.0.Co;2-c

Sinitcyn, P., Hamzeiy, H., Salinas Soto, F., Itzhak, D., McCarthy, F., Wichmann, C., ... Cox, J. (2021). MaxDIA enables library-based and library-free data-independent acquisition proteomics. *Nature Biotechnology*, *39*(12), 1563-1573. https://doi.org/10.1038/s41587-021-00968-7

Skowronek, P., Krohs, F., Lubeck, M., Wallmann, G., Itang, E. C. M., Koval, P., ... Mann, M. (2023). Synchro-PASEF Allows Precursor-Specific Fragment Ion Extraction and Interference Removal in Data-Independent Acquisition. *Molecular & Cellular Proteomics*, *22*(2). https://doi.org/10.1016/j.mcpro.2022.100489

Smith, P. K., Krohn, R. I., Hermanson, G. T., Mallia, A. K., Gartner, F. H., Provenzano, M. D., ... Klenk, D. C. (1985). Measurement of protein using bicinchoninic acid. *Analytical Biochemistry*, *150*(1), 76-85. https://doi.org/10.1016/0003-2697(85)90442-7

Sobott, F., Watt, S. J., Smith, J., Edelmann, M. J., Kramer, H. B., & Kessler, B. M. (2011). Comparison of CID versus ETD based MS/MS fragmentation for the analysis of protein ubiquitination. *Journal of the American Society for Mass Spectrometry*, *20*(9), 1652-1659. https://doi.org/10.1016/j.jasms.2009.04.023

Sun, Y., Xing, Z., Liang, S., Miao, Z., Zhuo, L.-b., Jiang, W., ... Guo, T. (2024). metaExpertPro: A Computational Workflow for Metaproteomics Spectral Library Construction and Data-Independent Acquisition Mass Spectrometry Data Analysis. *Molecular & Cellular Proteomics*, *23*(10). https://doi.org/10.1016/j.mcpro.2024.100840

Tsai, Y. S., Scherl, A., Shaw, J. L., MacKay, C. L., Shaffer, S. A., Langridge-Smith, P. R. R., & Goodlett, D. R. (2011). Precursor ion independent algorithm for top-down shotgun proteomics. *Journal of the American Society for Mass Spectrometry*, *20*(11), 2154-2166. https://doi.org/10.1016/j.jasms.2009.07.024

Wang, M., Jiang, L., Jian, R., Chan, J. Y., Liu, Q., Snyder, M. P., ... Luigi Martelli, P. (2021). RobNorm: model-based robust normalization method for labeled quantitative mass spectrometry proteomics data. *Bioinformatics*, *37*(6), 815-821. https://doi.org/10.1093/bioinformatics/btaa904

Wang, W.-Q., Jensen, O. N., Møller, I. M., Hebelstrup, K. H., & Rogowska-Wrzesinska, A. (2018). Evaluation of sample preparation methods for mass spectrometry-based proteomic analysis of barley leaves. *Plant Methods*, *14*(1). https://doi.org/10.1186/s13007-018-0341-4

Whiteaker, J. R., Sharma, K., Hoffman, M. A., Kuhn, E., Zhao, L., Cocco, A. R., ... Paulovich, A. G. (2021). Targeted mass-spectrometry-based assays enable multiplex quantification of receptor tyrosine kinase, MAP kinase, and AKT signaling. *Cell Reports Methods*, *1*(3). https://doi.org/10.1016/j.crmeth.2021.100015

Wiśniewski, J. R. (2018). Filter-Aided Sample Preparation for Proteome Analysis. In *Microbial Proteomics* (pp. 3-10). https://doi.org/10.1007/978-1-4939-8695-8_1

Wiśniewski, J. R. (2019). Filter Aided Sample Preparation – A tutorial. *Analytica Chimica Acta*, *1090*, 23-30. https://doi.org/10.1016/j.aca.2019.08.032

Wiśniewski, J. R., Friedrich, A., Keller, T., Mann, M., & Koepsell, H. (2014). The Impact of High-Fat Diet on Metabolism and Immune Defense in Small Intestine Mucosa. *Journal of Proteome Research*, *14*(1), 353-365. https://doi.org/10.1021/pr500833v

Wiśniewski, J. R., Zielinska, D. F., & Mann, M. (2011). Comparison of ultrafiltration units for proteomic and N-glycoproteomic analysis by the filter-aided sample preparation method. *Analytical Biochemistry*, *410*(2), 307-309. https://doi.org/10.1016/j.ab.2010.12.004

Xiaoshi, W., Simone, S., & Benjamin, A. G. (2015). Chapter 4 - Application of Mass Spectrometry in Translational Epigenetics. 55-78. https://doi.org/https://doi.org/10.1016/B978-0-12-801080-8.00004-1

Yang, K. L., Yu, F., Teo, G. C., Li, K., Demichev, V., Ralser, M., & Nesvizhskii, A. I. (2023). MSBooster: improving peptide identification rates using deep learning-based features. *Nature Communications*, *14*(1). https://doi.org/10.1038/s41467-023-40129-9

Yang, L., George, J., & Wang, J. (2019). Deep Profiling of Cellular Heterogeneity by Emerging Single-Cell Proteomic Technologies. *Proteomics*, *20*(13). https://doi.org/10.1002/pmic.201900226

Yen, C.-Y., Russell, S., Mendoza, A. M., Meyer-Arendt, K., Sun, S., Cios, K. J., ... Resing, K. A. (2006). Improving Sensitivity in Shotgun Proteomics Using a Peptide-Centric Database with Reduced Complexity: Protease Cleavage and SCX Elution Rules from

Data Mining of MS/MS Spectra. *Analytical Chemistry*, *78*(4), 1071-1084. https://doi.org/10.1021/ac051127f

Yoshida, H., Mizukoshi, T., Hirayama, K., & Miyano, H. (2006). On-line desalting–mass spectrometry system for the structural determination of hydrophilic metabolites, using a column switching technique and a volatile ion-pairing reagent. *Journal of Chromatography A*, *1119*(1-2), 315-321. https://doi.org/10.1016/j.chroma.2006.03.098

Yost, R. A., & Boyd, R. K. (1990). [7] Tandem mass spectrometry: Quadrupole and hybrid instruments. In *Mass Spectrometry* (pp. 154-200). https://doi.org/10.1016/0076-6879(90)93415-h

Yu, F., Teo, G. C., Kong, A. T., Fröhlich, K., Li, G. X., Demichev, V., & Nesvizhskii, A. I. (2023). Analysis of DIA proteomics data using MSFragger-DIA and FragPipe computational platform. *Nature Communications*, *14*(1). https://doi.org/10.1038/s41467-023-39869-5

Yu, F., Teo, G. C., Kong, A. T., Haynes, S. E., Avtonomov, D. M., Geiszler, D. J., & Nesvizhskii, A. I. (2020). Identification of modified peptides using localization-aware open search. *Nature Communications*, *11*(1). https://doi.org/10.1038/s41467-020-17921-y

Zecha, J., Satpathy, S., Kanashova, T., Avanessian, S. C., Kane, M. H., Clauser, K. R., ... Kuster, B. (2019). TMT Labeling for the Masses: A Robust and Cost-efficient, In-solution Labeling Approach. *Molecular & Cellular Proteomics*, *18*(7), 1468-1478. https://doi.org/10.1074/mcp.TIR119.
001385

Zhang, B., Whiteaker, J. R., Hoofnagle, A. N., Baird, G. S., Rodland, K. D., & Paulovich, A. G. (2018). Clinical potential of mass spectrometry-based proteogenomics. *Nature Reviews Clinical Oncology*, *16*(4), 256-268. https://doi.org/10.1038/s41571-018-0135-7

Zhang, F., Ge, W., Huang, L., Li, D., Liu, L., Dong, Z., ... Guo, T. (2023). A Comparative Analysis of Data Analysis Tools for Data-Independent Acquisition Mass Spectrometry. *Molecular & Cellular Proteomics*, *22*(9). https://doi.org/10.1016/j.mcpro.2023.100623

Zhang, Y., Sun, H., Zhang, J., Brasier, A. R., & Zhao, Y. (2017). Quantitative Assessment of the Effects of Trypsin Digestion Methods on Affinity Purification–Mass Spectrometry-based Protein–Protein Interaction Analysis. *Journal of Proteome Research*, *16*(8), 3068-3082. https://doi.org/10.1021/acs.jproteome.7b00432

Zielinska, D. F., Gnad, F., Jedrusik-Bode, M., Wiśniewski, J. R., & Mann, M. (2009). Caenorhabditis elegans Has a Phosphoproteome Atypical for Metazoans That Is Enriched in Developmental and Sex Determination Proteins. *Journal of Proteome Research*, *8*(8), 4039-4049. https://doi.org/10.1021/pr900384k

Chapter 3

A Morphological and Anatomical Evaluation of Common *Plantago* Taxa

Muzaffer Mükemre[1], PhD
Abdullah Dalar[2,3,*], PhD
and Murat Ünal[4], PhD

[1]Department of Plant and Animal Production, Yuksekova Vocational School, Hakkari University, Hakkari, Türkiye
[2]Department of Pharmaceutical Botany, Faculty of Pharmacy, Van Yuzuncu Yil University, Van, Türkiye
[3]Department of Medicinal and Aromatic Plants, Vocational Higher School of Kızıltepe, Mardin Artuklu University, Mardin, Türkiye
[4]Faculty of Education, Department of Biology Education, Van Yuzuncu Yil University, Van, Türkiye

Abstract

Raw materials of *Plantago* genera drugs are collected according to morphological similarity and are used for food and medical treatment purposes. These drugs are usually prepared in powder form, without given scientific names, and are sold in herbal markets. The plant material must be correctly identified for effective therapeutic use and treatment. This chapter aims to unveil the detailed morphology and anatomical structures of common *Plantago* species used as food and medicine and suggest scientific identification procedures both for plant and drug samples to contribute to the pharmacobotanical and systematical database of medicinal plants. Therefore, comparative pharmacobotanical characteristics of common members of the *Plantago* genus (*Plantago*

* Corresponding Author's Email: dalar.abdullah@gmail.com

In: Advances in Biology. Volume 9
Editor: Charles D. Grant
ISBN: 979-8-89530-574-4

anatolica Tutel & R. Mill., *Plantago atrata* Hoppe, *Plantago lanceolata* L., *Plantago major* L, and *Plantago media* L.) were investigated.

Morphologically, the leaves are lanceolate in *P. anatolica*, *P. lanceolata* and *P. atrata* species and elliptic-ovate in *P. major* and *P. media* species. Scapes of *P. anatolica* are distinctly 11-sulcate and glabrous; *P. lanceolata* is 5-sulcate and hairy; *P.atrata* is terete-round and densely hairy; *P. major* is slightly striate with l-celled hairs; and *P. media* is striate and densely septate hairy. Wide and emerginate bracts of *P. atrata* are a distinctive character. Spikes of *P. anatolica* and *P.major* are long (elongate); in *P. lanceolata* and *P. atrata*, they are globose; and in *P. media*, they are narrowly cylindrical. The fruit is a capsule in all taxa: cylindrical-ellipsoid in *P. anatolica*, conical in *P. lanceolata*, ovoid in *P. atrata*, ovoid-conical in *P. major* and ellipsoid in *P. media*. The structure of the seed is trigonous in *P.major* and cymbiform in other all species, with only *P. atrata* having a keel on the dorsal side. The surfaces of the seeds are alveolate in *P. anatolica*, smooth in *P. lanceolata*, and rugose in *P. atrata*, *P. major* and *P. media*.

Anatomically, there are no seperative anatomical structures between these taxa, with the exception of the presence of inulin crystals in *P. anatolica.* In the cross sections of the scape, it was clearly determined that the taxa were sulcate, striate or terete. When the leaf petiole cross section was examined, the shape of *P. anatolica* was a spiral, *P. atrata* was lunate, *P. lanceolata* and *P. media* were convex with 7 rounded corners and *P. major* was convex with 5 rounded corners. The number of vascular bundles is characteristic for *Plantago* species. The type of mesophyll was isolatheral in *P. anatolica, P. atrata* and *P. lanceolata* and dorsiventral in *P. major* and *P. media*. The type of stomata was found to be anomocytic and polocytic in *P. anatolica,* anomocytic and diacytic in *P. atrata*, diacytic in *P. lanceolata* and anomocytic in *P. major* and *P. media* in the superficial section of leaf. These morphological and anatomical features are important characters that facilitate the identification of the *Plantago* species used for food and medical treatment purposes. Based on these morphological and anatomical characters, novel key identification protocols were proposed for common *Plantago* taxa.

Keywords: anatomical, morphological, *Plantago,* identification key

Introduction

The *Plantaginaceae* family consists of 120 genera and 1,614 species, subspecies, and variety level taxa, and is dispersed in the high mountains of temperate and tropic regions. This family contains 217 taxa belonging to 18 genera in Türkiye (Güner et al., 2012). *Plantago* is among the cosmopolitan genus, as it contains 268 taxa worldwide. The genus name *Plantago* originates from the Latin word "planta," which means "sole of the foot," and refers to some taxa's big base leaves that touch the ground (Pilger, 1937).

In traditional medicine, *Plantago* is commonly used for various therapeutic purposes, as it has anti-bleeding, antimicrobial, anti-inflammatory, diuretic, expectorant, sedative and analgesic properties, and has treated gout, skin diseases and cancer cases worldwide (Samuelsen, 2000; Beara et al., 2012; Gonçalves and Romano, 2016). In Türkiye, *Plantago* species have traditionally been used for abdominal pain, hemorrhoids, injuries, abscesses, inflammation, ulcers, cancer, and diabetes (Tabata et al., 1994; Altundag and Ozturk, 2011; Kaval et al., 2014; Mükemre et al., 2015; Dalar et al., 2018; Dalar and Mükemre, 2020).

Plantago species have pronounced biological activities, such as high antioxidant, cytotoxic, anti-inflammatory, wound-healing, anti-obesity, and antiviral properties, which indicate their high therapeutic potential (Beara et al., 2012; Dalar and Konczak, 2013; Gonçalves and Romano, 2016; Mukemre et al., 2020). Apart from their pharmaceutical use, *Plantago* species are among one of the most regularly consumed vegetables globally. For instance, *P. asiatica* is a very popular vegetable in China for weight loss (Xu and Xu, 2007). The seeds and leaves of *P. coronopus, P. lanceolata*, and *P. serraria* are used for salads and baby food in France, Italy, and South Africa (Pieroni et al., 2002). *Plantago* species are consumed by people as well as grazing animals. Recent field surveys indicate a gradual decrease in their habitat due to early and over grazing. Field studies show that *Plantago* species cover an area of approximately ±50 km^2 (criterion B2a,b(iii)), with around 500 mature plants (criterion C).

Medicinal plants play an important role in the development of modern drugs. Approximately 25% of the current drugs used in modern Pharmacopoiea are directly or indirectly derived from plant materials (Sofowora, 1993; Pramono, 2002). Members of the *Plantago* genus have been widely used as food and medicine worldwide. In Türkiye, drugs obtained from the *Plantago* genus are often collected from the wild based on morphological similarities. These drugs are usually prepared in powder form without scientific names and are sold in herbal markets (aktar). Healers in Eastern Türkiye (locally called "Hekim") have extensive practical experience in collecting wild plants and applying traditional medical medicine but lack familiarity with scientific plant classification (species, subspecies, and varieties). They instead use local plant names based on morphological characteristics and habitat features, which can lead to misidentification due to morphological similarities. Therefore, scientific classification and detailed descriptions of botanicals used for medicinal purposes are urgently needed.

It should be noted that across various rural areas, no unified plant nomenclature used by healers exists. In other words, local plant names vary between villages. This inconsistency is a major factors in incorrect application. Plant material must be scientifically identified for effective use and treatment. Morphological and anatomical studies on medicinal plants are important for authenticity and the quality of raw drugs. Pharmacobotanical studies are very important for the reliability and quality control of herbal drugs. *Plantago* species are morphologically very similar; therefore their scientific identities can not easily be identified through their macroscopic features alone. Morphologically similar *Plantago* species have different phytochemical composition and therapeutic potential (Samuelsen, 2000; Garg et al. 2012). Thus, this chapter aims to present a detailed morphological and anatomical analysis of various organs, including root, scapus, leaf, flower, and seeds of common *Plantago* taxa (*P. lanceolata*, *P. major*, *P. anatolica*, *P. atrata*, and *P. media*) from a pharmabotanical perspective to reveal reliable identification keys for both plant species and drugs.

Materials and Methods

Plant samples *(P.anatolica, P.atrata*, *P.lanceolata*, *P.major* and *P.media*) were collected during the vegetation period (flowering, fruit, and seed stages) from different localities and similar habitats in Van Province. For morphological studies and herbarium samples, the specimens were dried according to standard herbarium techniques and stored at Hakkari University Biodiversity Research Herbarium (VPH), Hakkari, Türkiye. Plant materials were preserved in 70% alcohol, which was later used for anatomical studies (Figure 1).

Anatomical observations were carried out on root, scape, petiol, leaf transverse sections, and upper and lower surface sections of the leaves. All sections were prepared manually, stained with Lugol and liquid Aniline Blue solution (Figure 2). The photographs of the sections were taken using a Leica DM500 research microscope.

Figure 1. Plant samples preserved in alcohol (70%).

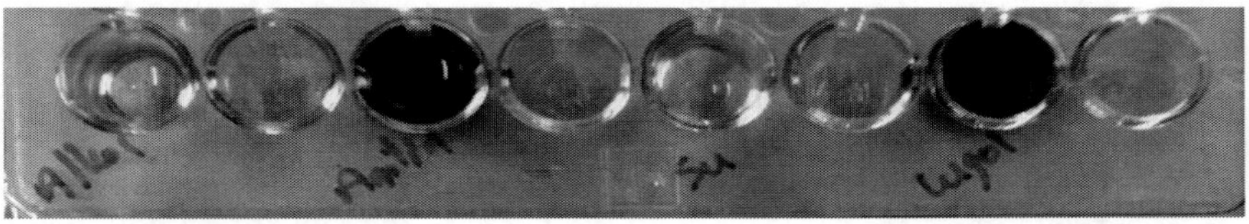

Figure 2. Dyes used for preparations.

The following steps were applied for the staining of the preparations:

a. Firstly, the plant material to be sectioned was cut into 1-2 cm pieces and kept in 96% ethanol solution for 5 minutes.
b. At the end of the period, the material removed from the alcohol was cut with a razor blade. To cut sections from plant materials, such as leaves or hairy roots, which are particularly sensitive, a piece of the material was cut and placed in Styrofoam, and thin sections were taken with a razor blade.
c. The sections were kept in aniline blue for 20-25 seconds.
d. The sections removed from aniline blue were first dipped into distilled water and then kept in Lugol solution for 20-25 seconds.
e. At the end of the period, the sections were removed from the Lugol solution and washed with distilled water. After placing the sections in a drop of pure water on a clean slide, a suitable coverslip was applied.
f. The prepared preparations were examined under a microscope and photographs were taken.

Results

Morphological Features

Organoleptic properties of a material belonging to medicinal and aromatic plants are characteristic. Analyses of organoleptic investigations include the five senses. The basis of these analyses are appearance, color, size, fracture surface, surface properties, texture, odor and taste. Organoleptic properties are important guidelines for the quality control of medicinal plant materials (WHO, 2011). *Plantago* species have similar morphological features and local names (Figure 3). They were collected from the same or nearby habitats and compared with several macroscopic and microscopic features for scientific identification (Table 1; Figures 4-7).

Figure 3. General view of P. anatolica, P.lanceolata, P.atrata, P.major and P.media.

Table 1. Macroscopic and microscopic comparison of taxa belonging to the *Plantago* genus

	P. anatolica	***P. lanceolata***	***P. atrata***	***P. major***	***P. media***
Life:	Perennial	Perennial	Perennial	Perennial	Perennial
Size:	30-45 cm	6-90 cm	3-30 cm	5-70 cm	15-45 cm
Leaves: (Figure 2)	Narrowly elliptic, 8-18 x 1.5-3.2 cm,7 veined, glabrous, petiole 6-7 cm.	Lanceolate, 7-42 x 0.4-5 cm, 3-7 veined, glabrous, pilose or villous.	Lanceolate, 3.5-20 x 0.2-0.7 cm,3-5 veined, glabrous or short pilose.	Elliptic-ovate, 3-40 x 1-12 cm, petiole often equalling lamina; 3-5 veined ssp. major or 5-9 veined ssp. intermedia	Elliptic-ovate, 4-17 x 1.5-3.5 cm, gradually narrowed into petiole, 3-9-veined, septate hairs.
Scapes (Figure 2)	25-45 cm, distinctly 11-sulcate, glabrous.	5-85 cm, 5 sulcate, dpressed-hairy.	4.5-14.5 cm, terete-round and densely hairy	3-30 cm, slightly striate, 1-celled hairs	25-30 cm, striate, septate hairs.
Spike (Figure 2)	Spikes elongate, 5.5-15.5 cm, narrowly cylindrical, lower flowers remote.	Spikes 0.5-8 cm, cylindrical, conical-cylindrical or globose.	Spike 1-2.5 cm, globose or ovoid.	Spike 4-40 cm, narrowly cylindrical, lax at base but otherwise dense.	Spike 4-15 cm, narrowly cylindrical.
Bracts (Figure 3)	Oblong-ovate with a narrow keel extending to apex.	Bracts shortly acute, ± concave, glabrous or hairy	Orbicular, emarginate 4.5 x 5-6 mm.	Ovate, longer than sepals, upper 1 mm, shorter than sepals, glabrous.	Narrowly ovate, 2-2.5 mm, fleshy keel, often glabrous.
Sepals (Figure 3)	Ovate with a narrow keel.	Orbicular-ovate, 2.5-3.5 mm.	Ovate, 3 mm.	2-2.5 mm, green, glabrous.	2.5 mm, obovate-elliptic, glabrous, strongly keeled.
Corolla (Figure 3)	Corolla 5 mm, lobes shortly acuminate.	Brownish corolla lobes 1.5-2.5 mm, trigonous-ovate.	White corolla lobes white, brown centre, 2-2.5 mm, membranous.	Corolla tube 2.3-3 mm, glabrous; lobes less than 1 mm, triangular.	White-hyaline, lanceolate-ovate, corolla lobes shining.

Table 1. (Continued)

	P. anatolica	*P. lanceolata*	*P. atrata*	*P. major*	*P. media*
Capsule (Figure 4)	Capsule cylindrical-ellipsoid.	Capsule conical	Capsule ovoid.	Capsule 2-4 mm, ovoid-conical	Capsule ellipsoid.
Seeds (Şekil 5.9;5.10)	Seeds 2-4, cymbiform, dorsal side alveolate.	seeds 2, cymbiform, narrowly ellipsoid to oblong, dorsal side smooth.	seeds 2, cymbiform, with keel on dorsal side, rugose.	seeds 6-30 (6-13 subsp. major; 14-30 subsp. intermedia) trigonous, dorsal side rugose	seeds usually 4, cymbiform, dorsal side striate- rugulose.
Flowering time	May-July	April-October	June-August	April-September	June-August
Habits	Water meadows	Roadsides, waste places nad streamsides.	Banks, pastures, steppe, alpine stony slopes	Stream- and riversides, rocky mountain slopes, marshes and fields.	Dry grassland and steppe
Height	1750-2200 m	1300-2200 m	2100-3500 m	0-2300m	1600-2300 m

Figure 4. Leaf, scapus and spike shapes of P. anatolica, P.lanceolata, P.atrata, P.major and P.media.

When Table 1 and Figures 4-7 are examined; the leaves are lanceolate in *P. anatolica*, *P. lanceolata* and *P. atrata* species and elliptic-ovate in *P. major* and *P. media* species. Scapes of *P. anatolica* are distinctly 11-sulcate and glabrous; *P. lanceolata* are 5-sulcate and hairy; *P.atrata* are terete-round and densely hairy; *P. major* are slightly striate with l-celled hairs and *P. media* are striate and densely septate hairy. Wide and emerginate bracts of *P. atrata* are a distinctive characteristic. Spikes of *P. anatolica* and *P.major* are long (elongate); in *P. lanceolata* and *P. atrata*, they are globose; and in *P. media*, they are narrowly cylindrical. The fruit is the form of a capsule in all taxa: cylindrical-ellipsoid in *P. anatolica,* conical in *P. lanceolata,* ovoid in *P. atrata,* ovoid-conical in *P. major* and ellipsoid in *P.* media. The structure of the seeds is trigonous in *P.major* and cymbiform in other all species, with the exception of a keel on the dorsal side in *P. atrata*. The surfaces of the seed are alveolate in *P. anatolica*, smooth in *P. lanceolata*, and rugose in *P. atrata*, *P. major* and *P. media*.

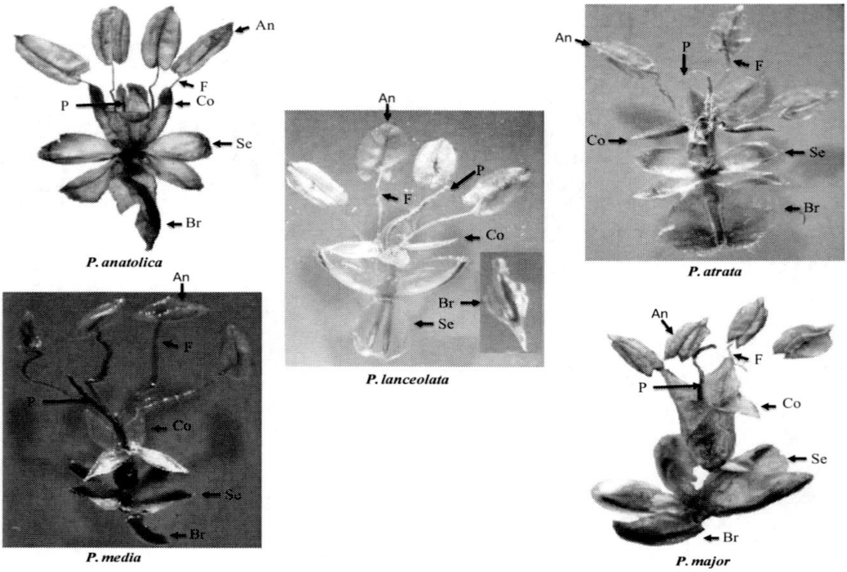

Figure 5. Flower structures of *P. anatolica, P.lanceolata, P.atrata, P.major* and *P.media.* Anther (An); Bract (Br); Corolla (C); Filament (F); Pistil (P); Sepal (Se).

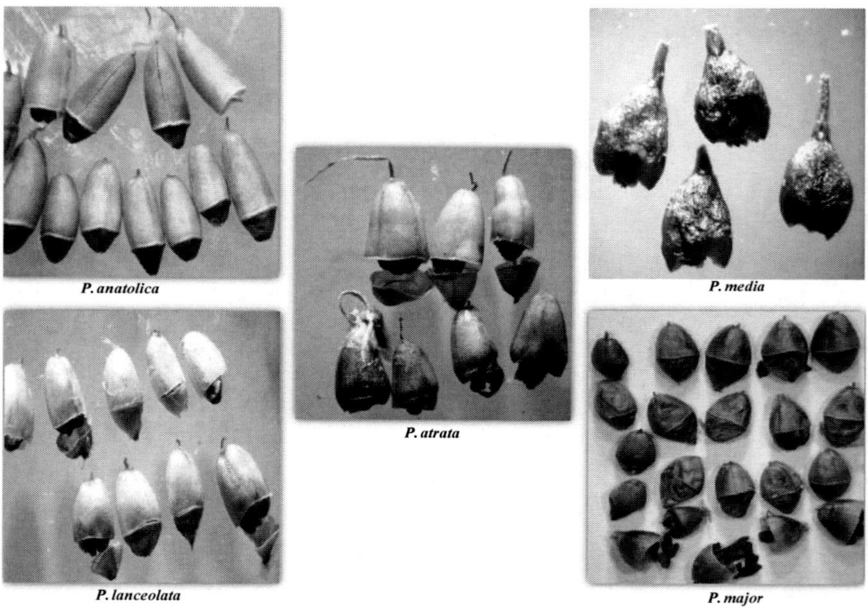

Figure 6. Capsules structures of P. anatolica, P.lanceolata, P.atrata, P.major and P. media.

Figure 7. Seeds structures of P. anatolica, P.lanceolata, P.atrata, P.major and P. media.

Anatomical Features

Root Anatomy

Anatomically, the sections taken from the root parts of the plants showed that there was no distinct anatomical structure between these taxa. However, inulin crystals were found only in *P. anatolica* (Figure 8).

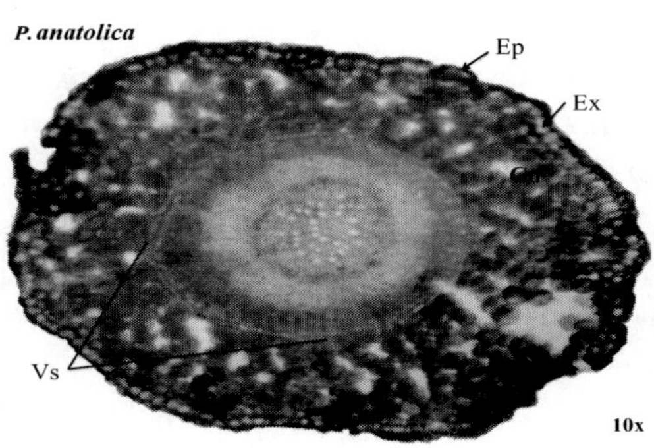

Figure 8. Cross section of roots for *P. anatolica, P.atrata*, *P.lanceolata*, *P.major* and *P.media*: Cortex (Co); Developing lateral root (Dv); Endodermis (En); Epidermis (Ep); Exodermis (Ex); Pith (Pi); Phloem (Ph); Sclerenchyma (Sc); Xylem (Xy); Vascular sylinder (Vs).

Scape Anatomy

Plantago anatolica Tutel & R. Mill.

Scape is typically 11-sulcate, and the epidermis consists of small, ovoid cells. Underneath the epidermis, there is a cortex consisting of ovoid parenchymatic cells. A single layer of endodermis (starch sheath) can be observed under the cortex. Sclerenchyma is located above the phloem under the endodermis. Vascular bundles are surrounded by an endodermis (starch sheath). The cambium is not distinguishable. The pith is present in the middle of the scape. Pith cells are round, vary in size, and have schizogen spaces between them. The middle part of the scapus is empty (Figure 9).

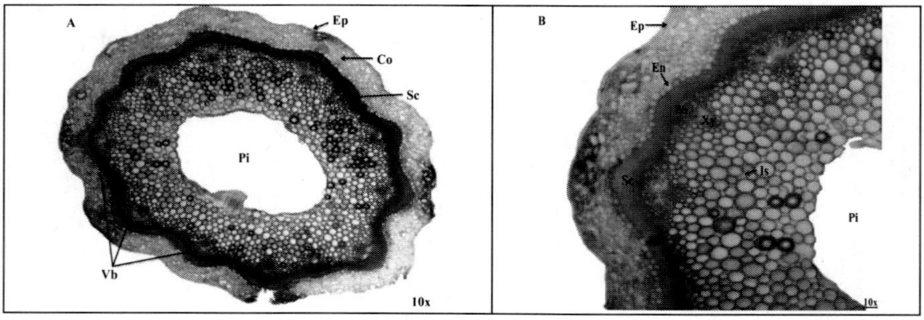

Figure 9. Cross section of the scape of *Plantago anatolica* Tutel & R. Mill. A. general view; B. detailed view. Cortex (Co); Endodermis (En); Epidermis (Ep); Intercellular space (Is); Pith (Pi); Phloem (Ph); Sclerenchyma (Sc); Xylem (Xy); Vascular bundle (Vb).

Plantago atrata Hoppe

The scape is terete and round shaped, and the epidermis is composed of rounded cells at the outermost side of the scapus. The cortex tissue consists of round and intercellular space parenchymatic cells. The single layer of the endodermis (starch sheath) can be observed under the cortex. There is plenty of starch in the endodermis. Sclerenchyma is located under the endodermis. Vascular bundles are surrounded by an endodermis (starch sheath). The cambium is not distinguishable. The pith is peresent in the middle of the scape and is completely composed of ovoid-orbicular parenchymatic cells (Figure 10).

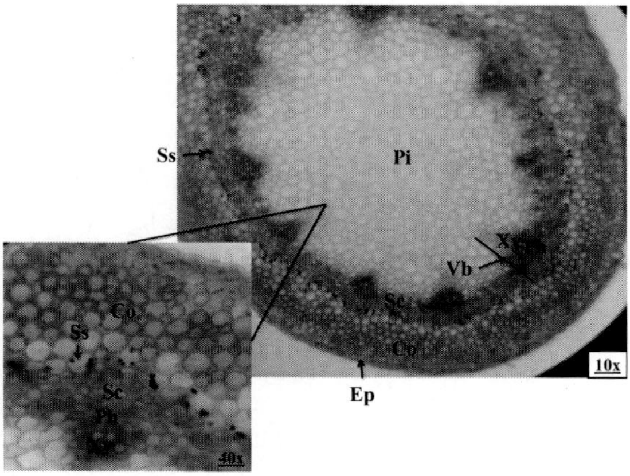

Figure 10. Cross section of the scape of *Plantago atrata* Hoppe. Cortex (Co); Endodermis (En); Epidermis (Ep); Pith (Pi); Phloem (Ph); Starch sheath (Ss); Sclerenchyma (Sc); Xylem (Xy); Vascular bundle (Vb).

Plantago lanceolata L.

The scape is typically 5-sulcate, which is clearly visible in the cross section of the scapus. The epidermis consists of a single layer, and its cells are small yet vary in ovoid shape. The cortex, consisting of a narrow layer of parenchymatic cells, is located under the epidermis. The endodermis consisting of flat cells is located between the epidermis and the sclerenchyma. The sclerenchyma layer has oval-round cells. Cambium is distinguishable. The pith is composed of large round parenchymatic cells. The middle part of the scapus is empty (Figure 11).

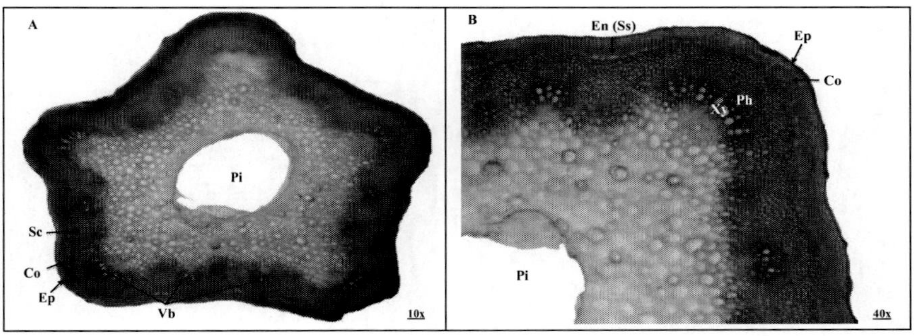

Figure 11. Cross section of the scape of *Plantago lanceolata* L. A. general view; B. detailed view. Cortex (Co); Endodermis (En); Epidermis (Ep); Pith (Pi); Phloem (Ph); Starch sheath (Ss); Sclerenchyma (Sc); Xylem (Xy); Vascular bundle (Vb).

Plantago major L.

The scape is slightly striate, the epidermis is a single layer, and its cells are small, with various orbicular shapes. Underneath the epidermis, there is cortex tissue consisting of parenchymatic cells. The single layer of endodermis (starch sheath) cam be observed under the cortex. There is a sclerenchyma layer undereath the endodermis. The vascular bundle and the pith region are wrapped around by the sclerenchyma layer. Cambium is not clearly distinguishable between phloem and xylem. The pith is peresent in the middle of the scape and is completely repleted by orbicular-ovoid parenchymatic cells (Figure 12).

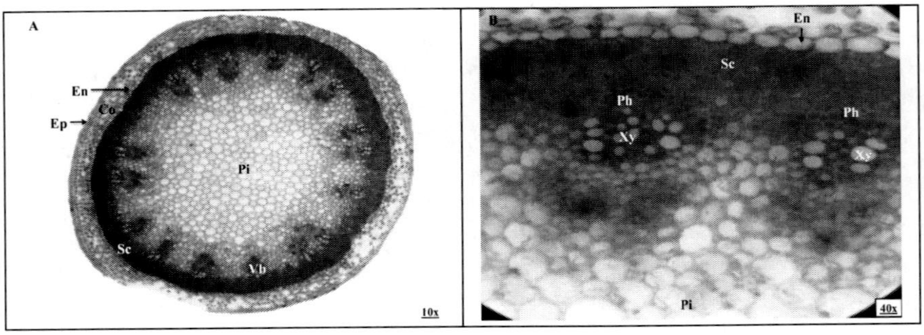

Figure 12. Cross section of the scape of *Plantago major* L. A. general view; B. detailed view. Cortex (Co); Endodermis (En); Epidermis (Ep); Pith (Pi); Phloem (Ph); Sclerenchyma (Sc); Xylem (Xy); Vascular bundle (Vb).

Plantago media L.

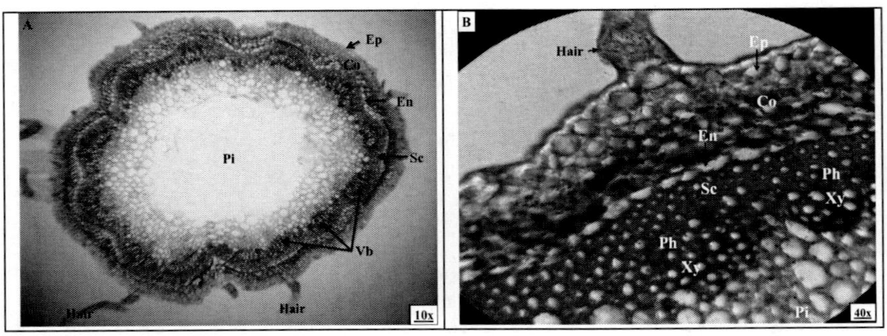

Figure 13. Cross section of the scape of *Plantago media* L. A. general view; B. detailed view. Cortex (Co); Endodermis (En); Epidermis (Ep); Pith (Pi); Phloem (Ph); Sclerenchyma (Sc); Xylem (Xy); Vascular bundle (Vb).

The scape is typically striate, with these striae clearly visible in the cross section of the scapus. The epidermis consists of various small, ovoid cells and is covered by septate hair. The cortex, consisting of ovoid parenchymatic cells, is located under the epidermis. Beneath the cortex, the single layer of the endodermis, consisting of regularly arranged ovoid cells, can be observed. The vascular bundle and the pith region are wrapped around by the sclerenchyma layer beneath the cortex tissue. The cambium is distinguishable. The pith, composed of orbicular and various sized cells, has intercellular spaces between them. The middle part of the scapus is empty (Figure 13).

Leaf Anatomy

Cross Section of Petiole

When examining the cross section of a petiole, the following shapes can be observed: *P. anatolica* is a spiral, *P. atrata* is lunate, *P. lanceolata* is convex with seven rounded corners, *P. major* is convex with five rounded corners, and *P. media* is convex with seven rounded corners and is also densely hairly (Figure 14).

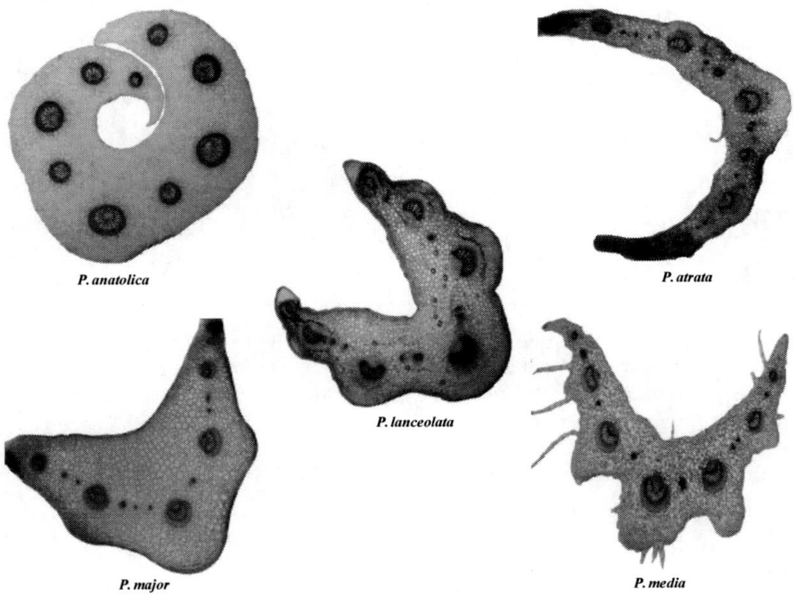

Figure 14. Cross sections of the petioles of *P. anatolica, P.atrata, P.lanceolata, P.major* and *P.media.*

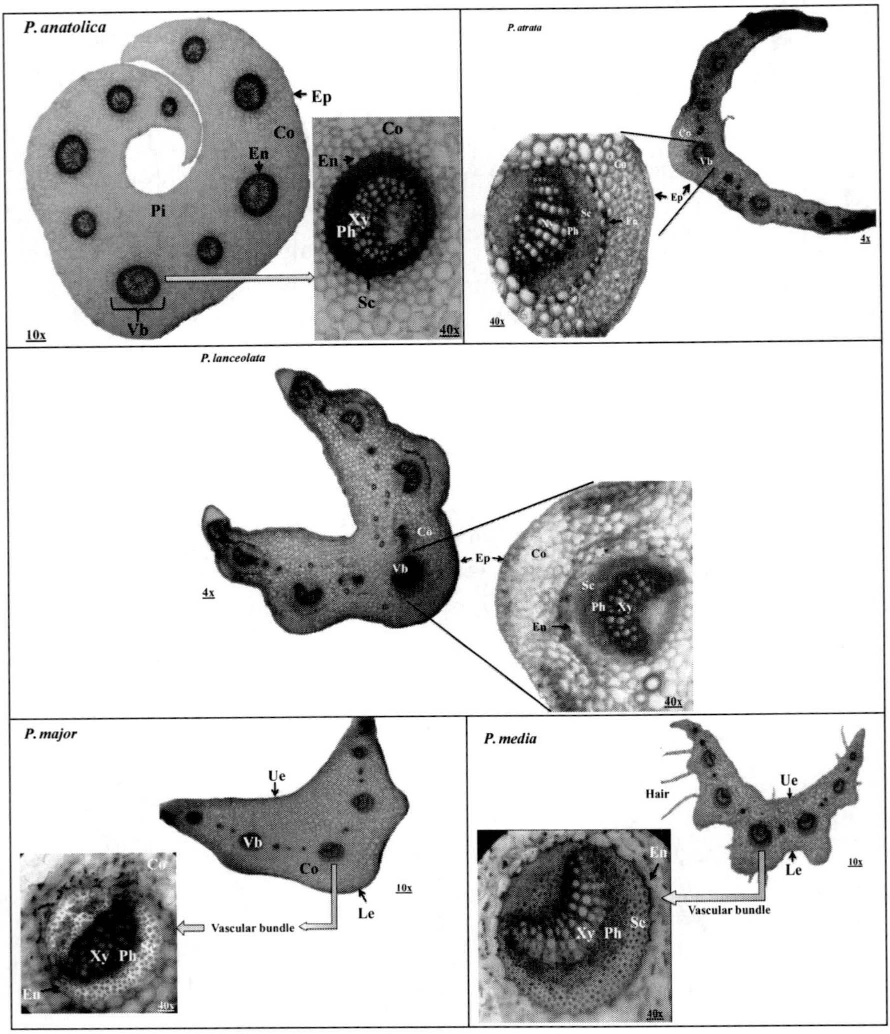

Figure 15. Vascular bundles embedded in parenchymatic cells in cross sections taken from the petioles of *P. anatolica, P.atrata*, *P.lanceolata*, *P.major* and *P.media.* Cortex (Co); Endodermis (En); Epidermis (Ep); Lower epidermis (Le); Pith (Pi); Phloem (Ph); Sclerenchyma (Sc); Upper epidermis (Up); Xylem (Xy); Vascular bundle (Vb).

In the cross section of the petiole, the number of vascular bundles embedded in the parenchymatic cells forming the cortex is characteristic for each taxa. In *P. anatolica*, there are seven advanced vascular bundles embedded in parenchymatic cells and two less advanced bundles on the leaf margin. *P. atrata* has three advanced and two less advanced bundles. *P.*

lanceolata and *P. media* have seven advanced bundles. The transmission bundles at the opposite corners are of equal size, while the vascular bundle at the most central oval corner is the largest. These anatomical properties support the morphological characteristics (Figure 15).

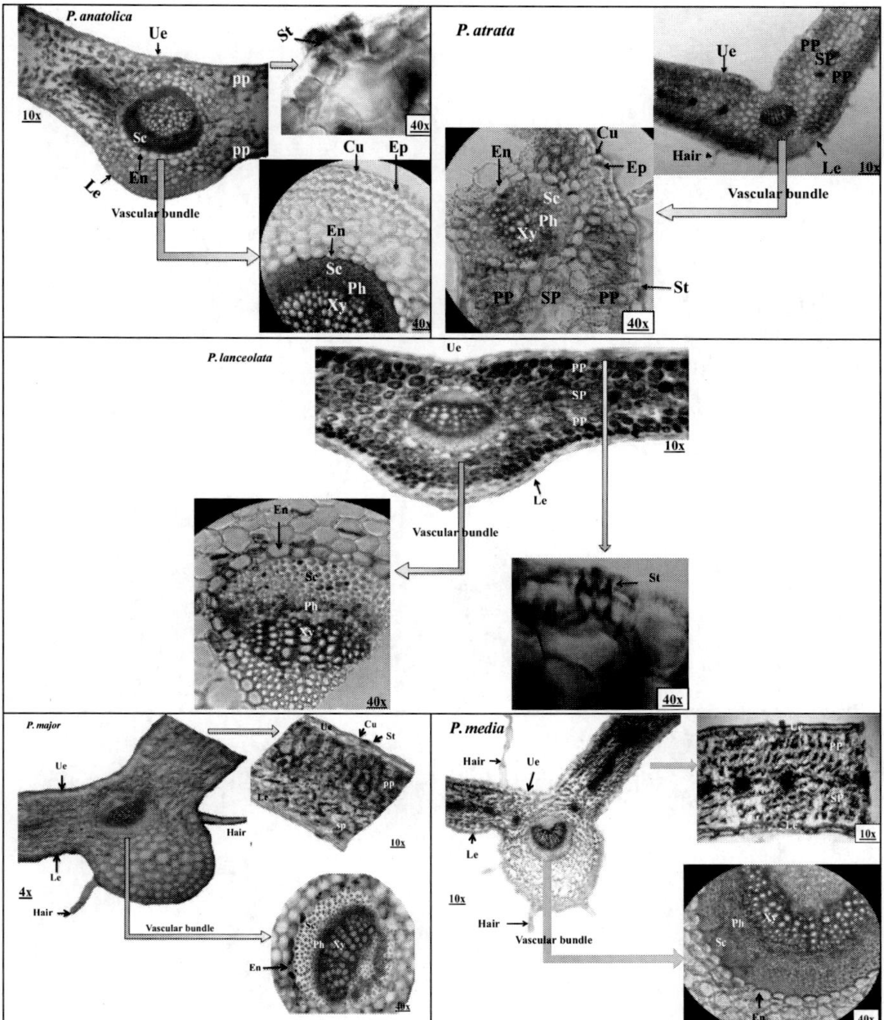

Figure 16. Transverse section of the central part of the leaf blade for *P. anatolica, P.atrata*, *P.lanceolata*, *P.major* and *P.media.* Cortex (Co); Cuticle (Cu); Endodermis (En); Epidermis (Ep); Lower epidermis (Le); palisade parenchyma (PP); Phloem (Ph); Sclerenchyma (Sc); spongy parenchyma (SP); Stomata (St); Upper epidermis (Up); Xylem (Xy); Vascular bundle (Vb).

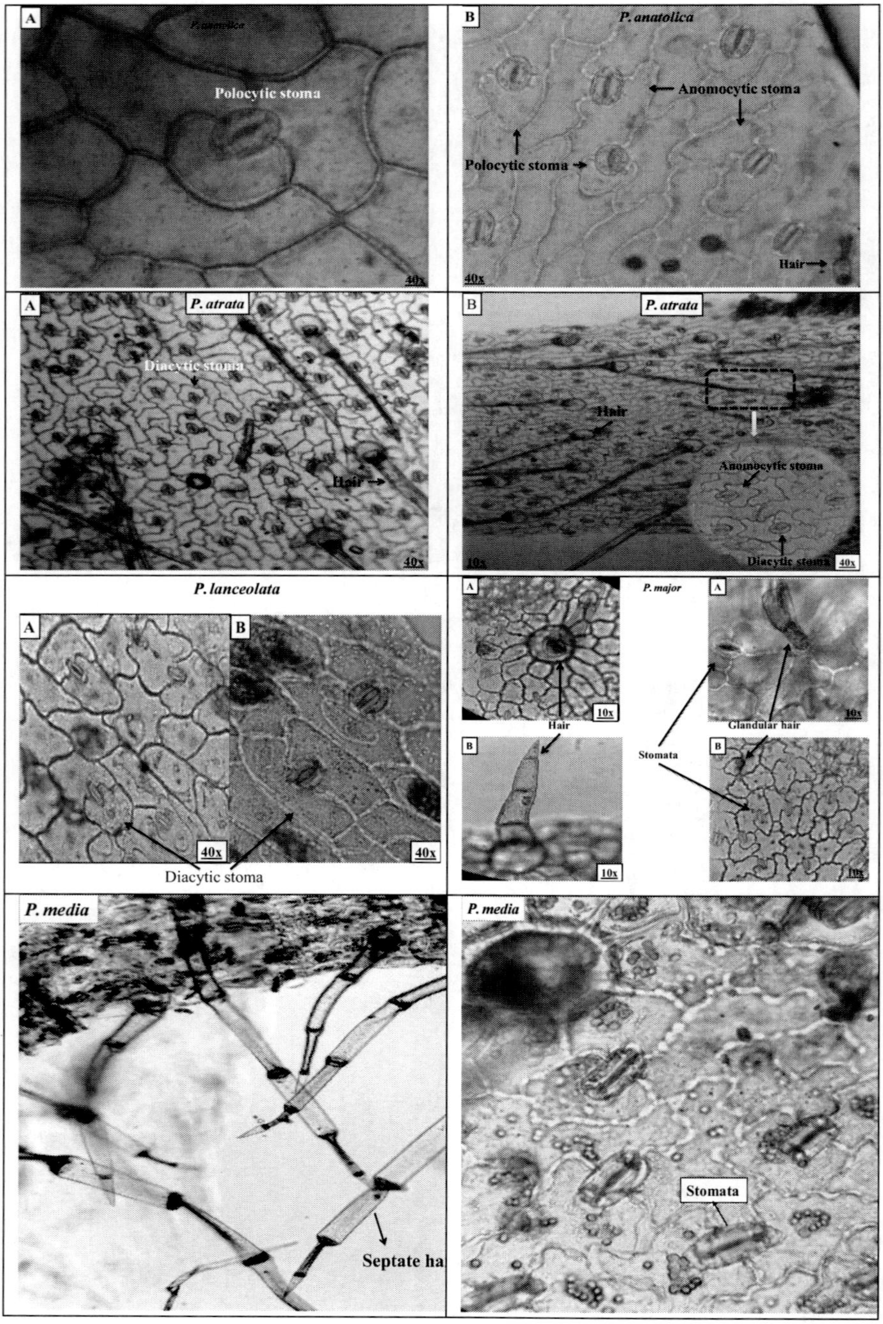

Figure 17. Superficial sections of the leaves *P. anatolica, P. atrata, P. lanceolata, P. major and P. media;* (A) Lower surface, (B) Upper surface.

Leaf Cross Sections

Based on the cross sections of *P. anatolica* and *P. major* leaves, they are considered hygromorphic because their stomas are located higher than the epidermis. The stomata in *P. atrata, P. lanceolata* and *P. media* show mesomorphic properties as they are at the same level as the epidermis. *P. anatolica, P. atrata* and *P. lanceolata* have isolateral mesophyll, with palisade parenchyma present on both sides of the spongy tissue. *P. major* and *P. media* have dorsiventral mesophyll, where the mesophyll is composed of palisade parenchyma near the upper epidermis and the spongy parenchyma cells in between the palisade and lower epidermis. The leaves of the *Plantago* species in this study are amphistomatic, meaning they have stomata on both the upper and lower surfaces of their leaves (Figure 16).

Superficial Sections in Leaves

In a superficial section of *P. anatolica*, the epidermal cell walls are glandular, and the stomata type is anomocytic and polocytic. Polocytic stomata is characterized by the solitary subsidiary cell of the stoma connected to its distal (marginopolar) side, where the subsidiary cell is U- or horseshoe-shaped (Van Cotthem, 1970). *Plantago atrata* is covered with single-cellular hair, and its stomata type is anomocytic and diacytic. In *Plantago lanceolata,* the epidermal cell walls are glandular, and the stomata type is diacytic. In *Plantago* major, the epidermal cell walls are hairy, and the stomata type is anomocytic. In *Plantago media*, the epidermal cell walls are more dense and septate-hairy, and the stomata type is anomocytic (Figure 17).

Discussion and Conclusion

The cosmopolitan *Plantago* genus includes 268 taxa worldwide. Species of *Plantago* are widely used in the treatment of various diseases, including bleeding, microbial infections, gout, skin diseases, diabetes, cancer, epilepsy, abdominal pain, hemorrhoids, injury, abscesses, inflammation, and ulcers (Samulsen, 2000; Tabata et al., 1994; Altundag and Ozturk, 2011; Beara et al., 2012; Kaval et al., 2014; Mükemre et al., 2015; Dalar et al., 2018; Dalar and Mukemre, 2020). Morphological and anatomical analyses of medicinal plants are important for contributing to pharmaceutical databases, preventing adulteration and improving the specific parameters the quality control of raw drugs (Bernerjee and Mukherjee, 2001). *Plantago* species have similar morphological characteristics, which can complicate precise identification. Specifically in taxonomic studies, minor anatomical differences are sometimes neglected. However, in fragmented plants (drugs), these anatomical differences can provide an advantage for correct and precise identification.

Sections taken from the root parts of *Plantago* taxa showed no distinctive anatomical structures among these taxa. However, inulin crystals were found only in *P. anatolica*, which can be considered a key characteristic. In cross sections of the scapus, *P. anatolica* has 11 sulcate and *P. lanceolata* has 5 sulcate; P. atrata is smooth, and *P. major* and *P. media* are striate. Although not very noticeable, abundant covering hairs appear from the epidermis layer of *P. atrata*, *P. lanceolata*, *P. major* and *P. media* plants (except *P. media*), while the absence of hairs on the scapus of *P. anatolica* plants is a distinguishing feature that supports the morphological characteristics of the taxa. Similarly, the number of vascular bundles embedded in the parenchymatic cells forming the cortex in the cross section of the leaf stalk (petiole) corresponds to the number of veins found in the leaves of the taxa, supporting the morphological character. Based on the anatomical data obtained, the anatomical features of the taxa belonging to the genus *Plantago* were given comparatively in Table 2.

Zou et al. (2001) reported that *Plantago* seed morphology is an extremely important characteristic for taxonomic investigations. Çiftçi (2020) examined the seed surface micromorphology of *P. atrata*, *P. lanceolata*, *P. major* and *P. media*, and reported that it was alveolate in *P. lanceolata* and rugose in other taxa. Çiftçi's results mirror the findings in this study in *P. atrata*, *P. major* and *P. media* taxa, but differ in *P. lanceolata* taxa. Our leaf anatomy findings agreed with Yoshida et al. (1997). In the superficial sections taken on the upper and lower surfaces of the leaf, anomocytic-type stoma was determined on the upper surface, and anisocytic-type stoma was determined on the lower surface of *P. major*; anomocytic-type stomata were detected on the upper and lower surfaces of *P. lanceolata* and *P. media*. Lanovici (2012) observed diacytic-type stoma in the superficial sections of *P. atrata*. In another study, Mesquita et al. (2017) observed anomocytic-type stomata in the superficial sections of *P. lanceolata* and *P. major*. In cross sections, it was reported that *P. lanceolata* had an isolateral-type mesophyll, and *P. major* had a dorsiventral-type mesophyll. Bahadar et al. (2018) reported anisocytic (paracytic) stomata on the upper and lower epidermis of *P. major* and *P. lanceolata*.

Table 2. Anatomical features of *Plantago* drug materials

	P. anatolica	*P. atrata*	*P. lanceolata*	*P. major*	*P. media*
Root	Inulin crystals present	Absent	Absent	Absent	Absent
Cross section of the scape	11-sulcate, glabrous	Terete-round, dense hairy	5-sulcate, hairy	Striate, hairy	Striate, septate hairy
Shape of the petiole in a cross section	Spiral	Lunate	convex with 7 rounded corners	convex with 5 rounded corners	convex with 7 rounded corners
Number of vascular bundles in petioles shown in a cross section	7 advanced and 2 less advanced vascular bundles	5 advanced and 2 less advanced vascular bundles	7 advanced vascular bundles	3 advanced and 2 less advanced vascular bundles	7 advanced vascular bundles
Type of stoma according to the arrangement of subsidary cells	Anomocytic, polocytic	Anomocytic, diacytic	Diacytic	Anomocytic	Anomocytic
Type of stoma according to the structure of the guard cells	Amaryllis	Amaryllis	Amaryllis	Amaryllis	Amaryllis
Type of stoma according to the plant's environment	Hygromorph	Mesomorpf	Mesomorph	Hygromorph	Mesomorph
Mesophyll type	Isolatheral	Isolatheral	Isolatheral	Dorsiventral	Dorsiventral

In conclusion, this chapter has presented detailed morphological and anatomical investigations of common *Plantago* species. Obtained data revealed an identification key for common *Plantago* species, which can be seen below.

Identification Key for the *Plantago* Species

1. Scapes prominently sulcate.
   2. Scapes distinctly 5- sulcate, hairy; leaves lanceolate, 3-7 veined; spikes cylindrical, conical or globose; seeds 2, cymbiform, narrowly ellipsoid to oblong, dorsal side smooth
      P. lanceolata
   2. Scapes distinctly 11-sulcate, glabrous; leaves narrowly elliptic, 7-veined; spike elongate; seeds 2-4, cymbiform, dorsal side alveolate.
      P. anatolica
1. Scapes terete and round, rarely striate.
   3. Leaves elliptic-ovate; bracts 2-3 mm, ovoid.
      4. Leaves elliptic-ovate; scapes with l-celled hairs, slightly striate; spikes 5-50 cm, narrowly cylindrical; seeds 6-30, trigonous, dorsal side rugose.
         P. major
      4. Leaves elliptic-ovate; shining white; scape septate hairy, slightly striate; spikes 3-15 cm; seeds usually 4, cymbiform, dorsal side striate- *rugulose.*
         P. media
   3. Leaves lanceolate; scape terete-round and dense hairy; bracts orbicular, 4-5 x 5-6 mm, emarginate at apex; seeds cymbiform, with keel on dorsal side, rugose.
      P. atrata

References

Altundag E, Ozturk M. Ethnomedicinal studies on the plant resources of east Anatolia, Turkey. *Procedia–Social and Behavioral Sciences* (2011) 19:756–777.

Beara I.N, Lesjak M.M, Orčić D.Z, Simin N.Đ, Četojević-Simin D.D, Božin B.N, Mimica-Dukić N.M. Comparative analysis of phenolic profile, antioxidant, anti-inflammatory and cytotoxic activity of two closely-related Plantain species: Plantago altissima L. and Plantago lanceolata L. *LWT – Food Science and Technology* (2012). 47(1):64–70.

Banerjee G, Mukherjee A. *Pharmacognostic studies on Portulaca oleracea L. I. Leaf.* In Maheshwari, J.K. and Jain, A.P. (Eds.) *Recent Researchs in Plant Anatomy and Morphology*. Scientific Publishers (India), JODHPUR; 2001.

Çiftçi A, Erol, O. Micromorphological studies on inflorescence and seeds of some plantain (Plantago L.) taxa in Turkey. *Biologica Nyssana* (2020) 11(2):85-91.

Dalar A, Konczak I. Phenolic contents, antioxidant capacities and inhibitory activities against key metabolic syndrome relevant enzymes of herbal teas from Eastern Anatolia. *Industrial Crops and Products* (2013) 44: 383-390.

Dalar A, Mukemre M, Unal M. Ozgokce F. Traditional medicinal plants of Ağrı Province, Turkey. *Journal of ethnopharmacology* (2018) 226:56–72.

Dalar A, Mükemre M. *Traditional Medicinal Plants of Van Province, Eastern Anatolia.* New-York: NOVA Science Publishers Inc. 2020.

Garg V, Dhar VJ, Sharma A, Dutt R. Facts about standardization of herbal medicine: a review. *Chin J Integr Med* (2012)10:1077–1083

Gonçalves S, Romano A. The medicinal potential of plants from the genus *Plantago* (Plantaginaceae). *Industrial Crops and Products* (2016) 83:213-226.

Güner A, Aslan S, Ekim T, Vural M, Babaç M.T, (edlr.). *Türkiye Bitkileri Listesi (Damarlı Bitkiler)*. Nezahat Gökyiğit Botanik Bahçesi ve Flora Araştırmaları Derneği Yayını. İstanbul. 2012.

Kaval İ, Behçet L, Çakılcıoglu U. Ethnobotanical study on medicinal plants in Geçitli and its surrounding (Hakkari-Turkey). *Journal of Ethnopharmacology* (2014) 155:171–184.

Lanovici N. Researches on Anatomical Adaptations of the Alpine Plants - *Plantago atrata. Annals of West University of Timişoara, ser. Biology* (2012) 15(1): 1-18.

Mesquita L.M.S, Colpo K.D, da Rocha C.Q, GattePicchi D, Tangerina M.M.P, Zachello-Nunes B, Martins M.B.G. Vilegas W. Anatomical differentiation and metabolomic profiling: A tool in the diagnostic characterization of some medicinal Plantago Species. *Revista Brasileira de Botanica* (2017) 40(3):801–810.

Mükemre M, Behçet L, Çakılcıoglu U. Ethnobotanical study on medicinal plants in villages of Çatak (Van-Turkey). *Journal of Ethnopharmacology* (2015) 166:361–374.

Mukemre M, Konczak I, Uzun Y, Dalar A. Phytochemical profile and biological activities of Anatolian Plantain (Plantago anatolica). *Food Bioscience* (2020) 36(3):100658

Pieroni A, Nebel S, Quave C, Münz H, Heinrich M. Ethnopharmacology of liakra: traditional weedy vegetables of the Arbëreshë of the Vulture area in southern Italy. *J. Ethnopharmacol* (2002) 81:165–185.

Pilger R. *Plantaginaceae.* In*: Engler A, ed. Das Pflanzenreich*, IV 269 (102, heft). Leipzig, Wilhelm Engelmann. 1937.

Pramono E. *The commercial use of Traditional knowledge and Medicinal Plants in Indonesia*. Multi-stakeholder Dialogue on Trade, *Intellectual Property and Biological Resources in Asia*, BRAC Rajendrapur, Bangladesh. (2002) 1–13.

Samuelsen A. B. The traditional uses, chemical constituents, and biological activities of Plantago major L. A review. *Journal of Ethnopharmacology* (2000) 71:1-21.

Sofowora A. *Medicinal Plants and Traditional Medicine in Africa,* Spectrum Books Limited, Ibadan, Nigeria. (1993):150.

Tabata M, Sezik E., Honda G, Yeşilada E, Fukui H, Goto K, Ikeshiro Y.

Traditional medicine in Turkey III. folk medicine in East Anatolia, Van and Bitlis provinces. *Pharmaceutical Biology* (1994) 32 (1):3-12.

Tutel B, Kandemir I, Kus S, Kence A. Classification of Turkish Plantago L. species using numerical taxonomy. *Turk J Bot* (2005) 29:51-61.

Xu G.S, Xu S.S. Development of composite health care beverage with plantain herb and caulis Lonicera japonica. *Food Sci.* (2007) 28: 616–619.

Van Cotthem W. R. J. A classification of stomatal types. *Bot. J. Linn. Soc.* (1970) 63:235-246.

World Health Organization (WHO). *Quality control methods for herbal materials.* WHO Press, Geneva, 2011.

Yoshida M, Tanaka T, Tsuboi H, Tabata M, Honda G, Sezik E, Yeşilada E. Studies on genus Plantago growing in Turkey (1) morphological comparison of leaf of the wild species. *Natural Medicines* (1997) 51(5):431-441.

Zou X, Foutain DW, Morgan ER. Anatomical and morphological studies of seed development in *Sandersonia aurantiaca* Hook. *S Afr J Bot* (2001) 67:183-92.

Chapter 4

Macrocyclic Ivermectin (IVM) as an Anti-Malarial Drug

Zohaib Saifi[1]
Afreen Inam[1]
and Rashid Ali[2,*]
[1]Medicinal Chemistry Laboratory, Department of Chemistry, Jamia Millia Islamia, New Delhi, India
[2]Organic and Supramolecular Functional Materials Research Laboratory, Department of Chemistry, Jamia Millia Islamia, Jamia Nagar, Okhla, New Delhi, India

Abstract

Ivermectin (IVM), a 16-membered macrolide in the family of avermectins (AVMs) for which its discovery – received the Nobel Prize (2015), continuously attracts global interest of scientists because of its widespread uses. Notably, its isolation *by* S. Omura through fermentation of the soil microorganism *Streptomyces avermitilis* has provided history for researchers. Since its approval by the FDA for the actions of crusted scabies as well as strongyloidiasis, it has become a very popular drug after the discovery of penicillin, another Nobel Prize-winning drug. It has gained recognition due to its many uses, including for COVID-19. Although the WHO did not recommend it for COVID patients during the pandemic and it is still under trial, it may open new doors for saving lives. Because of its remarkable benefits in treating various diseases, it continues to be widely studied. Here, we aimed to uncover its potential uses in treating malaria, and the authors are of the opinion that this

* Corresponding Author's Email: rali1@jmi.ac.in

In: Advances in Biology. Volume 9
Editor: Charles D. Grant
ISBN: 979-8-89530-574-4

wonder drug (IVM) will create another milestone for the sake of the future of mankind.

Keywords: Ivermectin (IVM), wonder drug, plasmodium, malaria, dosages, infection

1. Introduction

Malaria is a transferable but treatable and curable life-threatening disease, brought about by the plasmodium parasite. It is to be pointed out that male mosquito is not responsible for infection in humans, rather the egg-producing female Anopheles mosquitoes, are responsible for it. [1] The current well-known forms of plasmodium are *P. vivax, P. malariae, P. ovale, and P. knowlesi* and *P. falciparum,* - leading to malarial infection in humans. Notably, falciparum infection constitutes the prominent parasite which causes the most serious type of the condition, majorly contributing to the percentages of deaths and morbidity connected to the infection. The infection caused by *P. ovale* as well as *P. vivax* is minor because of the tendency of dormant hypnozoite formed in the liver. Noticeably, a non-infected mosquito is not capable of causing malaria. It is caused only by an infected female anopheles' mosquito which bites a human and in turn transmits the sporozoites of the parasite into the blood.

During the natural life cycle of malaria (Figure 1), both humans and Anopheles mosquitos become infected. The developed gametocytes are used to produce sporozoites by the female Anopheles mosquito vector. When an Anopheles mosquito bites in order to feed on blood, the developed sporozoites are transferred into the human host via the saliva of the female mosquito Anopheles. [2] Hepatocytes are infected by sporozoites, which then multiply and release merozoites into the circulation after invading liver cells. The merozoites subsequently undergo asexual reproduction in the erythrocytes, where they foster into the unripe ring-stage trophozoites and schizonts, respectively. Schizonts rupture and release merozoites into the circulation when they are fully developed. Blood stage-specific medications are essential for reducing symptoms of illness and fatality rates. The only parasitic phase that may spread from people to mosquito vectors is when certain merozoites change into reproductive erythrocytic processes that transform into immature gametocytes and are heralds of male and female gametes. [3] Then, following a blood meal, a mosquito with the species Anopheles consumes these gametocytes, which cause the mosquito to start reproducing. They eventually develop into sporozoites, which stay in mosquito glands that produce saliva, and are hungry to infect the new-fangled human host. [4] The natural *life cycle* of malaria parasites is presented in Figure 1.

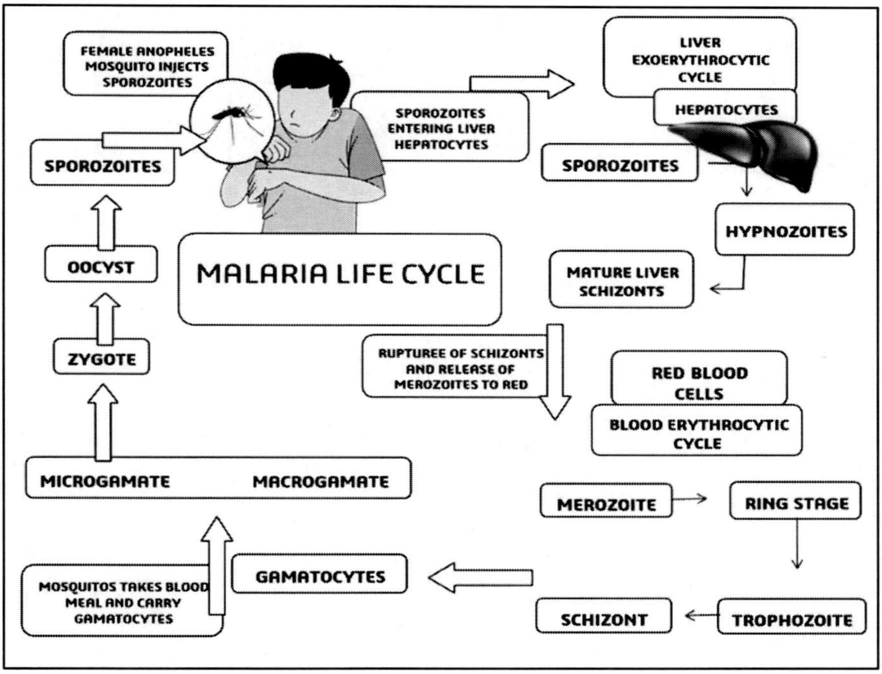

Figure 1. The natural life cycle of malaria parasites [5].

The WHO reported that a large part of the general public was at risk of infection in 2020. According to the latest data, approximately 241 million clinical cases of malaria and 0.627 million deaths worldwide were reported in 2020 in contrast to 227 million cases in 2019. This particular information shows about 14 million more clinical cases and 69000 more malaria deaths compared to 2019. This sudden jump in malaria cases and deaths in 2020 may be due to the COVID-19 pandemic disturbance. The graph in Figure 2 presents malaria-infected cases and death cases in 2010, and from 2017 to 2021. Malaria is a treatable disease if it is diagnosed in time and treated with prescribed medication. [1, 6] Preferred medications for pregnant women, adults and kids with malaria infection are ‘artemisinin combination therapy

(ACT)', and fast acting analogues of the artemisinin **1** with the retentive stand-in blood schizonticide. [7]

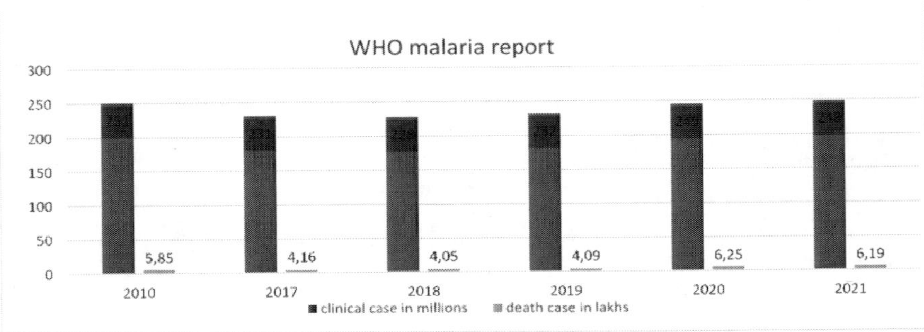

Figure 2. Bar graph representation of malaria report based on WHO (2010-2021).

The other category of blood schizonticidal prescription medication is designed to mitigate the chances of reemergence after the preliminary blood parasite load is cleared. [8] The most commonly implicated ACT-combinations are: primaquine **11**, pyrimethamine **10**, 4-sulfadoxine **9**, piperaquine **8**, dihydroartemisinin **7**, 4-mefloquine **6**, amodiaquine **5**, artesunate **4**, lumefantrine **3**, and artemether **2**, as displayed in the figure-3. Amongst them primaquine **11** is the most potent anti-malarial drug that has been resistive to the hypnozoite stages of the plasmodia. [1, 9] It is reasonably active against *P. falciparum gametocytes*, which makes it efficient for blocking the further advancement of the parasite. [10] The advance of defiance to the one-time front-line anti-malarial agents like chloroquine **13** and quinine **12**, sulfadoxine pyrimethamine, up-and-coming confrontation to the artemisinin [11, 12] and unavailability of the vaccine projects a grim image of malaria treatment. Researchers are working to develop anti-malarial drugs that target drug-resistant Plasmodium strains. These drugs are designed to be taken orally, work effectively with minimal side effects, and have new mechanisms of action.

According to the 'Global Technical Strategy for Malaria (GTSM) 2016-2030' [1], there is still a need for carrying out the advanced work toward reducing the significant risk of malaria infection, as current drug discovery and development processes are time consuming. Repurposing approved drugs for malaria can save time and offer benefits since their safety and effectiveness are already established.

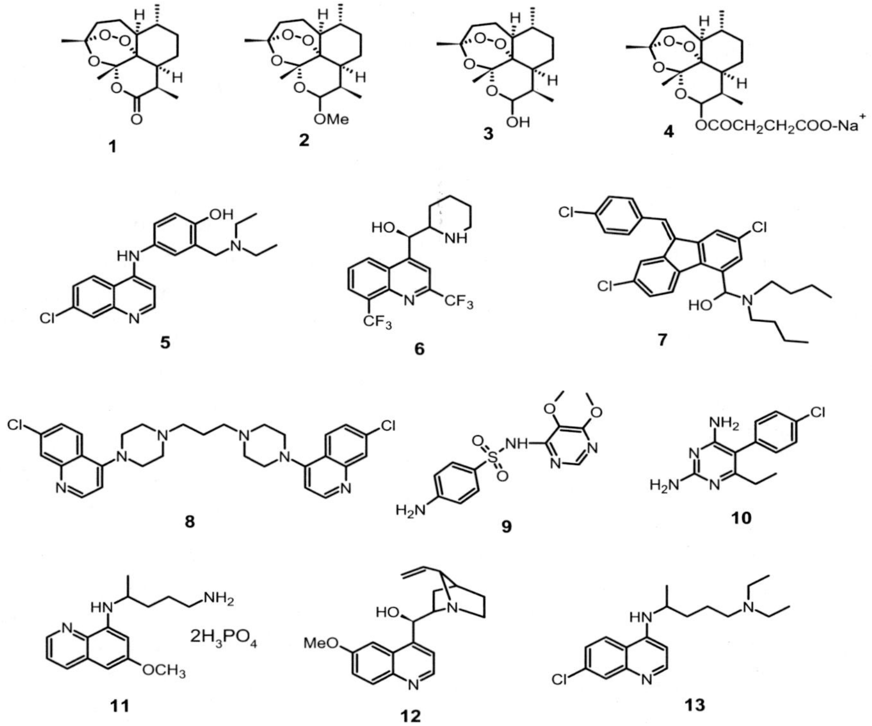

Figure 3. Structural representation of some important anti-malarial drugs.

As can be observed from scientific literature, the IVM is a semi-synthetic macrocyclic lactone containing drug molecule belonging to the avermectin (AVM) family – and extracted from the soil *actinomycete Streptomyces* avermitilus (Figure-4). IVM has a unique safety profile in mammals. According to broad research, it is used to reduce parasitic infection in humans as well as some other animals. Remarkably, Satoshi Omura secured the Nobel Prize in "*Medicine for the discovery of IVM drug*" in the year 2015. [13] Notably, IVM ranked as the 420th most commonly prescribed drug with more than one lakh prescriptions in the United States in 2018. IVM is a generic medicine found to be very effective against several ectoparasitic arthropods in addition to parasitic nematodes, and it was the first endectocide medicine in history, killing a wide range of parasites and vectors both within and outside the body. [14, 15]

In 1985, Ivermectin was proven in India to kill the main mosquito that spreads malaria. [16] Moreover, it proved to be an extremely innovative medicine that was unique in many aspects when it first arrived in the late

1970s. Before the development of this particular medicine, Mura, who was the director of the group of Antibiotics Research of Kitasato Institute established a cooperation with the US-based '*Merck MSD pharmaceutical*' business. The researchers of the Kitasato Institute separated organisms from soil sampling and performed the first *in vitro* assessment of bioactivity according to the conditions of the research settlement. The MSD laboratory received potential bioactive samples for *in-vivo* testing, and from that an effective and promising new molecule (IVM) was invented. This bioactivity was eventually determined to be caused by a novel molecule, given the name avermectin. Despite extensive research over many years, the Japanese microbe remains the only known source of avermectin..

IVM, an extraordinarily safe and most effective avermectin derivative, originated from a sample of soil in Japan. It was developed through a revolutionary global collaborative research relationship to discover a novel anti-parasitic treatment that was first introduced as a commercialized product at the beginning level for animals. This excellent drug molecule has been found to be effective towards a wide variety of parasites including mites, lice, gastrointestinal roundworms, lungworms and horn flies. [14, 17, 18] It was initially made available to the veterinary market and had a significant impact on veterinary medicine. It has also helped to control the mange (demodicosis) and dermatitis (eczema). Since then, IVM's annual veterinary revenue has exceeded one billion dollars, and the success of this drug has led to a sizable research organization for the development of numerous veterinary purposes. [14] It was evaluated early for interest in human vector-borne parasitic infections, including Onchocerca volvulus microfilaria.

The transmission-blocking outcome of IVM when administered in annual mass drug administrations (MDA) was tested in numerous medical trials carried out in endemic tropic nations for such diseases in the latter period of the 20th century. These have proven to be extremely safe with adverse events being limited to the mild '*Mazzoti reactions*' related to death microfilaria in severely contaminated patients. Women who are just starting to breastfeed are typically barred from this IVM-MDA due to restricted security data. Despite this, they are actually considered to be more secure than utilizing older drugs. They are even more effective than insect sprays used against blackflies and mosquitoes that spread the infection. The rare use of MDA with IVM has greatly reduced the number of disease cases. It may also help eliminate the disease from large areas in the future. As a result, onchocerciasis is no longer a problem in many parts of Central as well as South America besides in a few locations of Africa, however disposal of lymphatic filariasis (LF) are relatively slower to accomplish. [19-21] More importantly, IVM has been used to treat onchocerciasis infection legitimately throughout the course of recent years with the help of community-wide MDA of this drug adding to close to end of such infection. [22] IVM reduces the lifespan of Anopheles mosquitoes exposed to the drug - resulting drop-off the malaria transmission. We may get even better outcomes in coming years with proper dosing of the IVM or its combination with some other anti-malarial drugs, as taking higher doses sometimes causes side effects. [23-25] According to several reports, the IVM has enough potential to survive in the body's circulatory system for a long period. This helps ensure the correct drug levels to effectively eliminate the malaria parasite. [26-28]

2. Avermectins (AVMs)

Avermectins (AVMs) are classified as semisynthetic macrocyclic 16-membered lactones derived from the *Streptomyces Avermitilis* fermentation. The major and minor components have effectively been limited by chromatographic techniques. Two vital components of AVMs, that is, B1a and B1b represented in Figure 4, are actually the main content of the IVM drug. Over the years, AVMs have been used to treat a wide window of ectoparasitic and endoparasitic (*primarily parasitic nematode*) infections in animals as well as in humans. [13, 29-31] IVM is simply a dihydro derivative of AVM which consists of B1a > 80%, and B1b < 20%, obtained from the AVM (B1) by means of selective hydrogenation utilizing the Wilkinson's catalyst. [14, 32] As far as safety and efficacy are concerned, IVM was found to be even better than the original form of AVMs.

This potent drug was authorized in the early '80s as a medication in the veterinary field with good activity, and it has also been used in treating onchocerciasis, lymphatic filariasis, and a few other different conditions. Although IVM is a very effective drug, when performing MDA, some minor to mild side effects, like arthralgia, coma, headache, *etc*., have been reported in several patients. Remarkably, IVM has successfully been applied to humans. Nematode transmission is reduced by killing the contagious microfilaria phases and by limiting the growth as well as fertility of the adult worms in people. [31-35] A great deal of research has examined the negative effects of IVM on various stages of the malaria life cycle, including vector control. The neurological condition that results from the following are hyper-polarization of the neurons and muscle fibers which causes the insect's death.

[28] Glutamate-gated chloride (GluCl) channels are a subtype of "*ligand-gated ion channels*" predominantly found in invertebrates, specifically in parasites and insects. These vital channels play a crucial role in regulating the normal functioning of their nervous systems. Notably, GluCl channels are absent in the '*central nervous system*' (CNS) of humans. As a result, IVM, a drug that acts on GluCl channels, exhibits selective action on invertebrates, rendering it an effective and safe antiparasitic agent for humans, with minimal impact on our own nervous system. The lack of GluCl channels in humans further prevents IVM from crossing the blood-brain barrier, thereby limiting its influence on the CNS and reducing the likelihood of neurological side effects in humans. These fascinating highlights of the IVM can open up new door of expectations for using this wonder drug or its modified versions as option drugs in humans. [33, 34] A better drug candidate could be designed and constructed in the light of IVM in coming years, and the modified drug can limit the resistance or cross-resistance issues, permitting the execution of coordinated malaria.

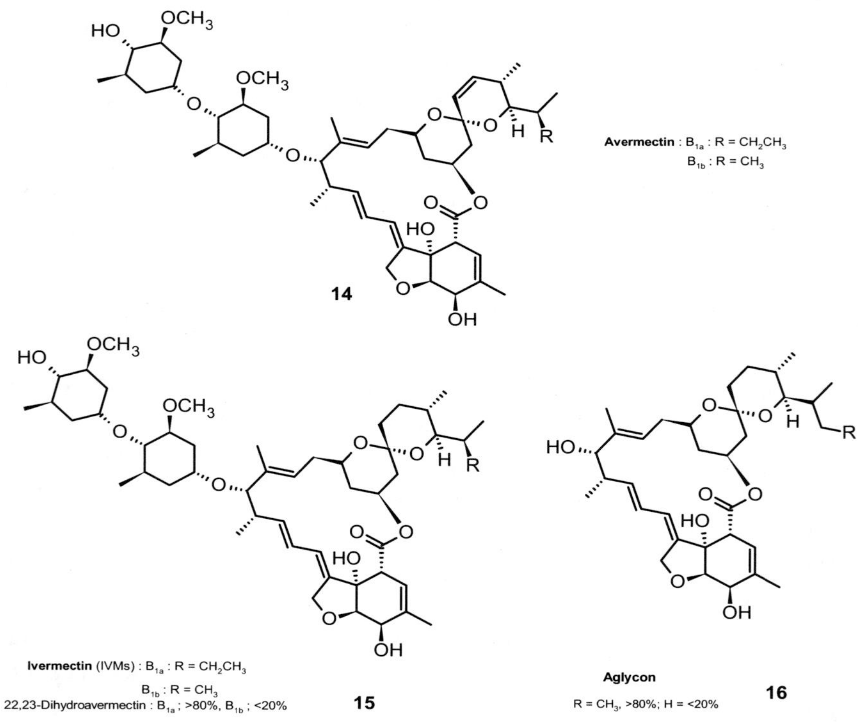

Figure 4. The structures of representative AVMs (**14**), IVMs (**15**), and aglycon (**16**).

2.1. Extraction Procedure, Validation Method and LC-MS Optimization of the Ivermectin

For the extraction of IVM, a variety of solvents have been extensively used, after optimizing the conditions, distilled H_2O:MeCN in 2:3 (v:v) was found to be the best combination. This particular mixture was also employed for the isolation of '*spiked-insect tissue samples*' at 0.6 ng/g and 6 ng/g fortification level. On the other hand, it was observed that all corroboration considerations directed that the procedure to identify IVM in the '*dung beetle tissues*' for living or dead beetles and the hemolymph, was vastly precise, reproducible, and sensible for appraisal of varied analytical methods. Notably, the samples which were spiked at the concentration levels for example 0.6 ng/g and 6 ng/g, unveiled a recovery from "91.1 ± 11.6% to 103.2 ± 6.4%", respectively. From these outcomes, it was concluded that after continuous SPE extraction, IVM can completely be extracted from the beetle tissue samples with negligible loss in the macrolides lactone during the complete procedure. Moreover, the concentration perceived by the instrument (3:1 signal/noise ratio), was inferred as the IVM limit of detection (LOD), which is valued to be 0.01 ng/g., while with 10:1 signal/noise ratio the limit of quantification (LOQ) was found to be 0.1 ng/g. Lastly, the calibration standards were also successfully run-in triplicate, displaying worthy linearity in the window of 0.1 ng ml–1 to 10 ng ml–1 (r = 0.988).

For the detection of IVM, the best LC-MS ionization condition was realized with the usage of "*direct infusion electrospray solvent*" such as acetonitrile/water ammonium formate buffer conducted in both positive as well as negative modes. However, the negative ion signals were found to be considerably lower in comparison to the positive ion signals at very low concentrations of the IVM drug. From the chromatographic separation, retention times for the IVM was noticed to be 10.1, and the entire run time for every specific injection was found to be less than 11 minutes. The IVM mass spectra in ESI positive mode displayed a fundamental peak ca. at *m/z* 897.4 from $[B1a{+}Na]^+$ in spite of the anticipated peak at *m/z* 875 $[B1a{+}H]^+$, along with three key daughter ions peaks at *m/z* = 753.3, 835.3 and 609.4 – showing the adducts with the sodium (Na) ion. These daughter ion peaks were used for the overall identification as well as quantification of the IVM drug. Finally, for the quantitative intends, instrument was functioned in the '*multiple reaction monitoring mode*' by means of the daughter ions *m/z* 607.3, 753.3, & 835.4) for the IVM. [36, 37]

3. Action of Ivermectin towards Malaria Parasite

Ivermectin reduces the lifespan of Anopheles mosquitoes exposed to the drug, resulting in a decrease in malaria transmission. For example, the endectocidal effect should be sustained sufficiently for a long period to restrict vector limit by giving higher doses, multiple doses, or using sustain-release doses. The adequacy should be adjusted against the gamble of secondary effects, which might increase at higher dosages. However, the efficacy should be adjusted, otherwise risk of side effects may be increased with higher doses. [24-25]

3.1. Ivermectin's Mechanism of Action

GluCl channels, the site of action of the IVM, by which synaptic transmissions are blocked in invertebrates, leads to the hyper-polarization, death of the mosquitoes and paralysis. The cys-loop class of "*ligand-gated ion channels*" is not completed without such GluCl channels. It is demonstrated that IVM influences a number of other members, including histamine, gamma-aminobutyric acid (GABA), in addition to the chloride channels. [38, 39] Noticeably, action of IVM in mammals is that of an allosteric $GABA_A$ receptor agonist. These GABA receptors are positioned in the CNS regions and enhance the chloride conductance, resultant in hyperpolarization as well as reduced formation of action potentials. [40-41] The GABA receptor is also a major inhibitory transmitter in vertebrates. Net upshot of the $GABA_A$ receptor stimulus is the "*central nervous depression*" (CND), which characterizes the condition of IVM drug poisonousness in vertebrates.

4. Is IVM Effective or Ineffective towards Malaria Parasites?

There isn't a clear description of how IVM affects the plasmodium parasites right away, hence detailed experimental as well as theoretical studies are of supreme demand. A new medication screening concentrates on at first exhibiting IVM's viability against hepatic *P. berghei.* Another recent study exhibiting some proof that IVM might restrain improvement of *P. falciparum* and *P. berghei.* [28] In a separate study, a more seasoned *in vitro* study with *P. falciparum* assessed that the IC_{50} as well as IC_{90} of IVM were extremely contrasted with blood plasma concentrations, resulting from the usage of a single dose (150 μg/kg) for the treatment of onchocerciasis. In this manner, recommended standard dose of IVM are probably not going to influence the *P. falciparum* advancement. [42] Based upon available literature reports, imaginable anti-plasmodium activities in humans are quite completely missing. According to another studied data, it was revealed that, there was no impact of malaria infection using IVM dose (200 µg/kg). [43] On the other hand, in a primary trial to know about the impact of IVM drug onto the plasmodium sporogony, Kobylinski et al., tainted a lab strain of *genus Anopheles gambiae* with a cultured *P. falciparum* parasites before or after consumption of a dose which was less than lethal dose of IVM drug. [44] When IVM was consumed continuously for 6 and 9 days following infection, the data confirms an impact on the pervasiveness of plasmodium condition in mosquitoes. Research proposed that IVM might possess a transmission blocking effect. In any case, there was no IVM effect on infection intensity (number of malaria parasite oocyst phases), implying that IVM may not have an instant impact against malaria parasite. In a clinical trial based data, the authors have reported that there was no result of IVM on the sporogony after consumption of an infected blood including IVM and ACT. [45] In a different work, the authors have revealed lacking an IVM prophylactic result on *P. falciparum.* [46] According to these findings there are differences in methodologies, and IVM has no significant impact on the plasmodium parasites in mosquitos or merely a difference in mosquito qualities like vector competence and survival that are related to the ability to transmit disease. It is important to note that the ecology of malaria is quite variable, and the effects shown in one mosquito/parasite combination may change or could be ineffective in another five plasmodium species that causes human malaria and are being spread out through a variety of anopheles species. Natural plasmodium-anopheles species combinations are exceedingly particular, resulting in a unique epidemiology and transmission patterns. According to the modelling and different studies, the impact of IVM on malaria transmission is primarily driven by its mosquitocidal impact, but this requires it to be confirmed actively throughout the various malaria ecologies. [47]

5. Is the Action of Ivermectin on the Vector Stage Effective or Not?

Basically, vectors play an important role for lowering parasite transmission and avoiding malaria infection. Due to its effectiveness, vector management may become a crucial tool for malaria control and its elimination method. The likelihood of a mosquito surviving depends upon the host preference index, the interval between two repeated blood meals, and the vector density in relation to the host, are all the decisive factors in malaria. [48] Accordingly, membrane feeding experiments with *Aedes albopictus* and *Aedes aegypti* species using human blood laced with different doses of IVM (0.001–100 g/mL) demonstrated that after 7 days of the blood meal, all three animals had 100% death at the high dose values (10–100 g/mL). In addition, it was discovered that mosquitoes that survived sub-lethal dosages of IVM (0.001–0.1 g/mL) laid fewer eggs and hatched them less frequently. It is interesting that infertility could be reversed because normal egg production could resume after refeeding on the pure blood. [49] One distinguishing quality of the IVM that needs to be highlighted is the switching of fertility. Interestingly, many studies have demonstrated the vulnerability of adult mosquito species to MDA with the usage of IVM administered via different delivery routes and doses. [27, 50-51]

The survival of *Anopheles gambiae species* was shown to be reduced for up to six days following MDA (150 μg/kg) in an intriguing investigation carried out in Senegal. [50] Similar results were observed when different anopheles such as farauti mosquitoes were given permission to feed on a participant who had already received a single dose of the IVM (250 μg/kg), as only up to 20% of the mosquitoes might survive for three days. [51] On the other hand, the rate of mortality has been recorded as high in mosquitoes that ingested within the first week, and the mosquitocidal effect continued for at least 44 days in human blood, demonstrating a long-lasting residual effect of the IVM. *Anopheles gambiae* mosquitoes was injected on a volunteer group in a different trial who had either received IVM (200 μg/kg) a day before or nothing at all. [27] The mean survival time in the former group was 2.3 days as opposed to 5.5 days. However, there was no disparity in mortality between the two groups of mosquitoes fed after two weeks since the treatment.

A one-time dose of IVM (200 μg/kg) demonstrated important mosquitocidal activity against anopheles mosquito in an independent trial on MDA, carried out in three different villages in the Solomon Islands to prevent the spread of scabies. [52] A deadly outcome and its proficiency as a preferred strategy to mark parasite transmission in scabies endemic regions are indicated by the fact that roughly 80% of the mosquitoes caught on day 2 after MDA started dying throughout the period but only 75% of those caught on the day 1 after MDA did. Additionally, it is anticipated that IVM will provide double benefits in the cases of coexisting malaria parasitic illnesses, like lymphatic filariasis and onchocerciasis, because it may reduce the death and disease brought on by vector-borne/parasitic diseases. These investigations supported the effectiveness of the IVM drug as an insecticide in lowering vector survivability *via* MDA. Impact of IVM onto the vector (*A. gambiae*) populations in a laboratory environment and in the field was probed in an independent investigation. [53-54] The LC_{50} for *A. gambiae* was found to be 19.8 ppb, 9 days after consuming bovine blood spiked with IVM (0.011000 ppb). In this report, approximately 90% of *A. gambiae* mosquitoes permitted to feed directly on the cattle fed with IVM (600 μg/kg, injectable form) were unable to survive for more than 10 days after the blood feeding, and the impact persisted for 2 weeks after therapy. [55] The mosquito survival rate stayed below the measured value again till 20 days just after cattle treatment despite the fact that the mosquitocidal effects began to diminish over the time.

5.1. Effect of Combinations of Ivermectin on the Vector

Reduced mosquito survival was seen in a field investigation, designed to assess the effects of MDA using a combination of IVM (400 μg/kg), and diethylcarbamazine citrate (6 mg/kg) on the continued existence of anthropophagic anopheles punctulatus mosquitos, malaria parasite vectors in Papua New Guinea. Seventy percent of the blood-fed mosquitoes that were caught to a lesser extent than 4 days after MDA, all perished within 9 days. Additionally, when the 48h survival rate of A. punctulatus swollen with blood from those receiving treatment with IVM (400 μg/kg diethylcarbamazine citrate not included) was compared to that of those receiving no remedy, the former disclosed a significantly lower survival rate of 31% and latter 94%, outlining the influence of MDA. [53] It is therefore logical to expect that, similar to IVM, endectocides that are frequently utilized to regulate nematodes may be poisonous to malaria infected mosquito, and those who nourish onto the plasma of a host who has received any of these medications. Furthermore, with a good sense of therapeutic actions against vectors, malaria ecologies, and the tactical and routine delivery of such treatments, considerable influence over malaria parasite transfer could be predicted.

5.2. The Effect on Sporogony of Malaria Parasites

A report discussed in this chapter has conclusively shown that IVM, when consumed *via* blood meal, lowers *A. gambiae* and other species' survivability in both lab and field conditions. In a separate study, it was also found that decrease in the percentage of *P. falciparum* sporozoite encompassing *A. gambiae*, implying that IVM could inhibit *P. falciparum* sporogony at concentration range given in humans after receiving therapy that seems to be sublethal to *A. gambiae.* [50] IVM also had an impact on *P. falciparum* sporogony when IVM (1.07 mg/mL) and *P. falciparum* gametocytes were administered simultaneously (dose 1), or when IVM were given after 3, 6, or 9 days of (infected) blood meal (dose 2, 3, 4, respectively) as the equal dosages. The prevalence of oocysts was determined by dissecting mosquitos 7 days after a blood meal. Similarly, mosquitoes were dissected 12/14 days after the blood meal to assess the occurrence of sporozoites. While oocyst and sporozoite prevalence were both reduced at dose 1, sporozoite prevalence was reduced at dose 3 and dose 4. However, medication IVM showed no effect at dose 2 on the predominance of the two parasite types. No negative effects on the parasite's sporogony were noticed when the concentration of IVM in the blood level decreased up to 6.1 ng/mL. However, oocyst and sporozoite prevalence was shown to be lower when IVM (6.1 ng/mL) was given 3 days prior to the infected blood meal. [56]

5.3. Ivermectin Acting as Target Carriers of Malaria Parasites

Some parasites can stay in a single-cell form, called hypnozoites, inside human liver cells. They remain inactive for decades before possibly becoming active again. This makes *Plasmodium vivax* infection a major concern as it may return even after an effective therapy. [57] The cohabitation of *P. vivax* and *P. falciparum* infection in prevalent regions may make situation even more difficult. Primaquine (PQ) and tafenoquine are two commercially available 8-aminoquinolinebased medications to treat the hypnozoites or liver-stage infection. However, they are associated to intravascular hemolysis in individuals with glucose-6-phosphate dehydrogenase (G6PD) deficiency and are not advised for usage during pregnancy. [58-59] Additionally, *P. vivax's* genetic diversity is greater than that of other *Plasmodium species*, making malaria elimination strategies more effective. [60] Therefore, utilizing the medications that target the lever stage of malaria, a radical cure, that is total eradication of unexpressed liver-stage parasites (hypnozoites) observed in *P. vivax* infection) is quite possible. Anopheles vectors of several species including *Anopheles darlingi* and *Anopheles aquasalis* are capable of transmitting the *P. vivax* infection. [61-62] The survivorship of *A. darlingi* was found to be reduced in a study investigating the impact of IVM on time to refeed, *P. vivax* development, and *A. darlingi* survival (LC_{50} = 43.2 ng/mL, LC_{25} = 27.8 ng/mL, LC_5 = 14.82 ng/mL). Additionally, at concentrations corresponding to LC_{50} and LC_{25}, IVM was detrimental to *P. vivax* sporogony in *A. darlingi,* and reduced its prevalence to varying degrees (22.6 and 17.1 percent, respectively). [63] However, the '*average number*' of oocysts per mosquito fed onto the same blood source, known as oocyst intensity, at the above-mentioned values remained unchanged. After a single oral dose of IVM (200 μg/kg) in humans, it was observed that the levels dropped to 48.7, 26.9, 10.6, and 6.3 ng/mL afterward 4, 12, 36, and 60 hours, respectively. [63] As a result, *A. darlingi* were reduced in survival (LC_{50} = 43.2 ng/mL, LC_{25} = 27.8 ng/mL, LC_5 = 14.82 ng/mL) to evaluate the impact of IVM on the time to refeed. Importantly, IVM drug also reduced the prevalence of *P. vivax* in *A. darlingi* to various concentrations (22.6 and 17.1 percent, respectively) while maintaining the same oocyst intensity at the two concentrations corresponding to LC_{50} and LC_{25}. At the higher dosages (48.7 ng/mL and 26.9 ng/mL), the time to refeed was significantly decreased while at lesser concentrations, the effect was found to be negligible. [63]

5.4. The Effect of Ivermectin ACTs (Artemisinin-Based Combination Therapy) on Malaria Parasites

Only the asexual and early stages of gametocyte progress are negatively impacted by ACTs; mature gametocytes suffer lesser damage, making infected people more susceptible to malaria transmission. [64-65] Development of artemisinin resistance further complicates the problem, screaming for the generation of other methods for both treating and preventing the spread of residual malaria. In collaboration with ACTs partners, IVM could be envisioned as a promising drug for the eradication of malaria because of its effectiveness against the malaria vector population. Due to mosquitocidal effects of the IVM, it can block malaria transmission and can also supplement the benefit of ACT in removing immature gametocytes as well as asexual malaria parasites. This section presents sample studies that use IVM in combination with ACTs. On 120 asymptomatic *P. falciparum* carriers, a double blind, placebo-controlled trial was carried out by administering the time-dependent dosage of a mixture of lumefantrine (120 mg) and artemether (20 mg) (LA), (ii) LA combined with a single dose of IVM (200 μg/kg) or a placebo, and (iii) LA combined with a repeated dose of IVM (200 g/10-day survival assessment of *A. funestus* and *A. gambiae* on days 1, 3, and 7 after treatment with IVM and subsequent mosquito membrane feeding demonstrated a 4–7 fold rise in the death rate of mosquito. Additionally, during the first week of therapy, transmission was reduced by 27% for single-dose combinations and 35% for repeated dosage combinations, respectively. Furthermore, negligible effect on the death rate in mosquitoes fed on day 7 following treatments suggested that death was caused by the plasma concentration of the IVM, which subsequently dropped over the time. Furthermore, due to greater sustenance of IVM in plasma in the former linked to the frequently higher body mass index (BMI) of females, the mosquitocidal action of IVM declined slower in female patients in comparison to male patients. [65-68]

5.5. Ivermectin's Impact on the Vector

Controlling malaria transmission has an appealing feature of focusing on the reproductive fitness of the malaria vectors. The mosquitocidal effectiveness of IVM drug and the resulting effect on vector's ability to reproduce may have a significant impact on the vector population. The average number of eggs laid per female (22.44 in the LC_5 group vs. 34.70 in the control group) and the egg hatch rate (74.76 in the LC_5 group vs. 81.91 in the control group) significantly decreased when *A. aquasalis* mosquitoes fed on blood containing a sublethal concentration of the IVM (18.28 ng/mL), corresponding to LC_5 determined as the residual concentration after the ingestion of a blood meal. [69] These findings are consistent with a related study completed on cattle in Tanzania in which *A. arabiensis* feeding on the blood of cattle treated with IVM (200 μg/kg) caused 82% decrease in egg oviposition (the percentage of mosquitoes that laid eggs) and a 79% decrease in the fecundity in comparison to control assay. [70] After taking blood, more than 80% of the mosquitoes perished within 2–4 days, preventing them from finishing sporogony and producing eggs. The survival and reproductive health of *A. arabiensis* feeding on individuals treated with IVM was also found to be negatively impacted by the IVM drug. [71] Relative to the untreated controls, there was a decrease in egg production and mean surviving time. For the sake of objectivity and conciseness, this study will not go into detail to a number of other reports on the impact of IVM on the reproductive health of anopheles species. However, these data clearly demonstrate the effectiveness of IVM on the reproductive health of malaria vectors.. [46, 72-73]

6. Dosing and Administration

The dose of IVM drug is different for different diseases, ranging from 150-400 μg/kg. The mosquitocidal concentration lasting 48–72 hours with the single permitted dose of 200 μg/kg IVM in blood meal is equivalent to the LC_{50} in the instance of *A. gambiae.* [67, 69, 74] Multiple doses are therefore necessary to continue the mosquitocidal action at lethal or sublethal concentrations for a prolonged period. As a result, slow-release articulations and combination regimens with other medications that lengthen the half-life of IVM, investigated to address the logistical and operational issues faced in MDA. In one such investigation, rodents were intravenously administered with formulations of IVM comprising deoxycholate and sucrose utilizing Si-based implants in order to maintain, for 12 and 24 weeks respectively, blood concentrations of IVM corresponding to LC_{50} (10 and 7 ng/mL). [75-77] It was anticipated that there would be a considerable drop in the vector survivorship and a decline in a plenty of infectious bites that people would sustain. The author showed that IVM about 20 ng/mL may be maintained for 12 weeks in plasma in the larger mammals like pigs. [75] This helped maintain the ideal concentration needed to kill 50% of *A. gambiae* in 10 days. In a different study, performed on cattle, five Si-based slow-release implants, each containing about 73 mg of IVM, were used to maintain blood levels of IVM for more than 6 months - above 6 ng/mL, after subcutaneous administration. This concentration corresponds to the level required to cause 50% mortality in *A. gambiae* in 10 days of nourishing acquaintance. *A. arabiensis* mosquitoes were also evaluated for their 3-day survivability, and for more than 24 weeks, a fall in the survivability of about 50% was seen. The mosquito's inability to complete their gonotropic cycle and lay eggs, this may have a considerable effect on the transmission of residual malaria. [78-79]

7. The Effect of Ivermectin on Human Host Stages

Cytosolic, universally conserved ribonucleo-protein, called the signal recognition particle(SRP), detects and directs the transport of particular proteins to the endoplasmic reticulum, where they are effectively folded into the right 3D shape. [80] The transporter proteins karyopherin α (Kary α) and karyopherin β (Kary β), go through dimerization before attaching to SRP and moving *via* the "*nuclear pore complex*", assisting SRP's nucleocytoplasmic shuttling in *P. falciparum*. [28] The IVM drug restricts parasite growth by preventing SRP nucleocytoplasmic shuttling by either preventing Kary α and Kary β dimerization or trying to interfere with the SRP attachment to transporter the proteins. The drug was effectively antagonistic towards blood stages of *P. falciparum* infection, having *in-vitro* IC_{50} values of 2.85, 1.92, and 1.56 M, respectively against CQ-resistant (CQR) Dd2, M Camp strains and CQ-sensitive (CQS) 3d7 strains. Additionally, it has also been revealed that *in vivo* study on *P. berghei* infected mice showed that the parasitemia was suppressed by 40% compared to placebo-treated controls.

Moreover, IVM increased the host survival, as cured-mice outlived for 23 days in comparison to 10 days in placebo-treated mice. [28] The authors have used both *in-vivo* as well as *in-vitro* experiments to demonstrate the effectiveness of AVMs in expectation of hepatic-stage malaria infection. *In-vitro*, the drug IVM reduced *P. berghei* infection in human '*hepatoma cells*' with an efficacy than that of the commonly used liver-stage drug primaquine. IVM also impacted on host survival, as 80 percent of the preserved mice endured for 10-days as compared to 100 percent mortality in untreated mice during the same time period. The liver infection was reduced to 80 percent using IVM by 44–46 h after treatment to the mice. [81] These results suggested that IVM inhibits liver parasite development, which delay in the emergence of the blood-stage parasitemia. However, in a study that focused on a single stage, moderate *in vitro* suppression of liver-stage schizonts (IC_{50} = 10.42 M) and hypnozoites was observed in plasmodium cynomolgi-infected rhesus macaque hepatocytes when IVM was given prophylactically. [82] Furthermore, neither IVM alone (0.3, 0.6, or 0.12 mg/kg) nor in conjunction with chloroquine (10 mg/kg) prevented or delayed blood-stage parasitemia.

IVM also demonstrated a beneficial inhibitory effect on *P. falciparum's* asexual (IC_{50} = 100 nM) and sexual (IC_{50} = 500 nM) stages *in vitro*. A more recent in vivo study found that IVM had no effect on parasitemia or gametocytemia. [83] In addition, a human trial designed to assess direct prophylactic action of IVM revealed that a dose of 400 μg/kg given to volunteers two hours prior to the inoculation of *P. falciparum* sporozoites, had no effect on disease prophylaxis. The ambiguity in IVM's mode of action against parasite stages in human hosts was highlighted by these disputing through both *in vitro* as well as *in vivo* studies, highlighting the need for additional study to fully unravel the precise mechanism of action. [83-84]

7.1. Side Effects and the Emergence of Resistance

Humans can safely take a therapeutic amount of the IVM drug. [85] However, an excessive dose (more than 100 times the recommended amount) could have fatal results, such as coma. [86] Although, there was little evidence of serious negative effects in expecting mothers there is not enough information available to draw a connection between IVM and related adverse events. [87] Additionally, repeated MDA of the IVM (150–200 μg/kg) with a single dose of albendazole (400 mg) resulted in a 20% mitigation in clinical malaria incidents in children of maximum 5 years for 18 weeks. This occurred without any appreciable negative effects onto the general population, demonstrating the tolerability of repetitive dosing of IVM. [88] On the other hand, no CNS toxicity was observed in healthy adults even when treated with IVM at a 10-fold higher dose than that permitted by the FDA. Similarly, it was observed safe with onchocerciasis patients, after dosing as high as 1600 μg/kg to eliminate the parasite load. [89] These studies clearly confirm the safety of IVM in people, even using more than the advised doses.

Noticeably, very few reports are available in scientific literature about IVM's side effects. In the areas of Congo where onchocerciasis and L. loa are co-endemic, for instance, a retrospective investigation showed the prevalence of severe neurological as well as non-neurological side effects with MDA of IVM. These included severe headaches, coma, palpebral subconjunctival hemorrhages and arthralgia. [90] A study was carried out on four patients co-infected with *Onchocerca volvulus* and *L. loa* - showed side effects after MDA of IVM. They were assessed for mutations of the MDR1 gene. This was done to determine whether the IVM drug may express CNS toxicity because of its affinity toward GABA-Cls. [91-92] For an acceptable mosquitocidal effects, additional dosages and dosing frequencies are recommended. For the time being, two different dose regimens are being utilized in clinical trials to determine the safety and efficacy of IVM in humans during the time of the year when malaria is at its peak. During malaria season; a) the single dosage of IVM (400 μg/kg per month); and b) three consecutive doses of IVM (300

μg/kg per day continued every month) in combination with ACTs. While the former's safety and efficacy have been demonstrated in clinical trials, demonstrating no serious negative effects, the latter's safety and efficacy have only been established in a single clinical trial, and more clinical trials are required to reach a conclusion. [34]

Most of the dose's regimens developed for the effective MDA are dependent on the affected population's weight or height. This creates operational challenges, including variations in the number of people treated and incomplete treatment due to a higher risk of underdosing.. The shelter analysis disclosed that normal 18 and 36 mg dose regimens were well-tolerated, and found to be crystal safe, with the exception of the observation of the incidence of headache, when fixed high doses of IVM in the form of pills were prescribed to healthy patients rather than the reference-weight-dependent normal dosing strategy. [93] The probability of the emergence of resistance has grown as a result of the consistent prophylactic use of IVM against parasitic diseases. [94] To create improved anti-malarial medicines that work through various modes of action and can target various stages of the malaria life cycle, it is necessary to ascertain prospective indicators of resistance and accordingly design modified derivatives of the IVM drug to get promising molecular architectures.

8. Hybrids of Ivermectin

The biological effects of IVM are among the promising prospects for further research in the context of treating and eliminating malaria and are extensively covered in the sections above. For concision and relevance to the topic, this section focuses mostly on the structural alterations of the IVM in the light of anti-plasmodial activity.

Chemical changes resulting in the derivatization of the IVM drug that have been investigated for additional anti-parasitic and anti-helmintic characteristics are briefly presented only for comparison. The hybrids compounds **17-20** are inspired by the potency of the IVM against a part of its life cycle of *plasmodium species* including both *in-vivo* and *in-vitro*. The goal is to develop anti-malarial treatments with the potential to target multiple stages of the parasite's life cycle. [28, 94] These hybrid molecules were prepared and exhibited *in-vitro* activity against *P. falciparum* infection at the blood stage and *P. berghei* infection at the liver stage. The anti-plasmodial activity of IVM (IC_{50} = 2.1 μM *P. berghei*, 1.56μM against CQS NF54 strain) and its hybrid molecules from **17-20** are as (IC_{50} = 0.911 μM *P. berghei*. 161.2 nM against CQS NF54 strain), (IC_{50} = 0.503 μM *P. berghei* , 50.2 nM against CQS NF54 strain), (IC_{50} = 0.990 μM *P. berghei* , 110.5 nM against CQS NF54 strain) and (IC_{50} = 0.990 μM *P. berghei* , 256.7 nM against CQS NF54 strain) respectively. The most effective molecular hybrid, **18**, demonstrated three times greater activity than IVM against *P. berghei* hepatic-stage infection (IC_{50} = 0.503 μM) and eleven times better than *P. falciparum* erythrocytic stage infection (50.2 nM against CQS NF54 strain).

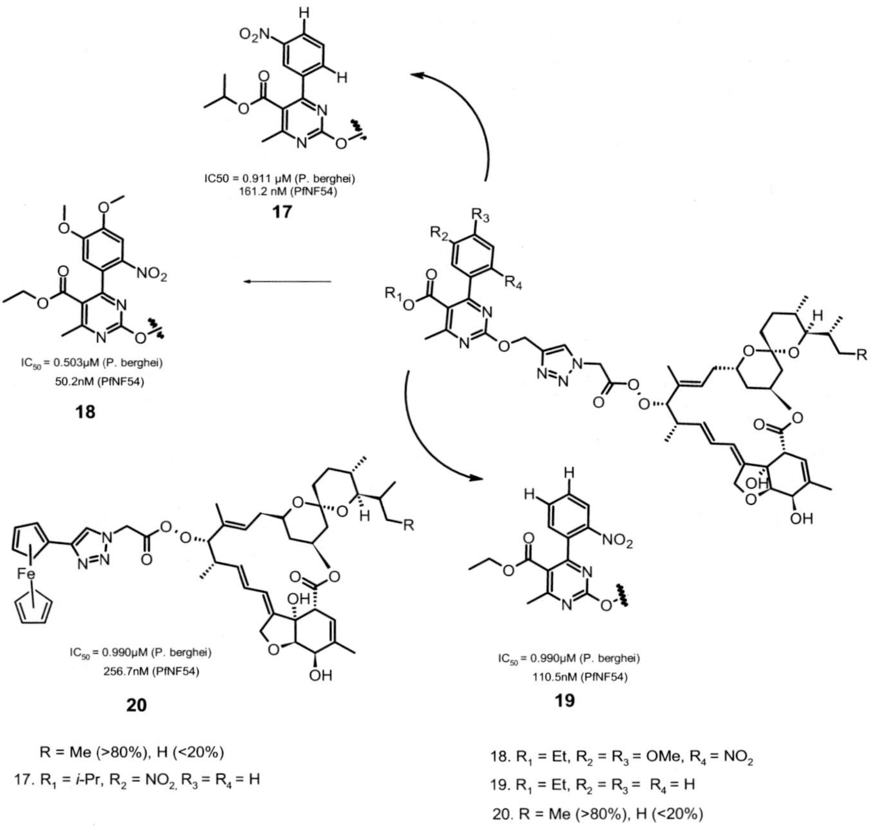

Figure 5. IVERMECTIN and its hybrid molecules with their anti-plasmodial activity.

Conclusion

Since the discovery of IVM by S. Omura in Japan, it is gaining more and more interest from the scientist community day by day because of its varied promising applications. Earlier, IVM was used in humans for treating the worm infestations for instance ascariasis, strongyloidiasis, enterobiasis, trichuriasis *etc*. However, during the COVID-19 pandemic, it gained significant attention because of its off-label usage for prophylaxis as well as in the treatment of COVID-19, but it is still under trial. In this account, the authors have exposed the anti-malarial uses of this Nobel Prize winning 'wonder drug' and discussed several cases with specific examples dealing with its effect, and dosing. In spite of these efforts to maximize IVM's advantages, more human trials still need to be conducted. The research community needs to be innovative and attentive in endeavoring to purge the world's most disabling, invasive, and stigmatizing parasitic infections. We believe that this drug will also be used in future for other diseases as well.

Acknowledgements

Dr. Afreen Inam is thankful to UGC-New Delhi for providing research funding. While Dr. Rashid Ali thanks DST-SERB New Delhi for financial support (Project File no. ECR/2017/000821). Zohaib Saifi is thankful to UGC New Delhi for providing Senior Research Fellowship (UGC Ref. No.: 108/(CSIR-UGC NET DEC. 2017). The authors are also highly thankful to Jamia Millia Islamia, New Delhi, India for stipulating the necessary research facilities.

Declarations

Ethical Approval

The authors declare no human and/or animal studies and we have followed ethical guidelines.

Consent for Publication

All the authors give consent for participation and publication.

Conflict of Interest

Authors declare no conflict of interests.

References

[1] WHO. World malaria report 2020: 20 years of global progress and challenges. *World malaria report*, 1-151. (2020).

[2] Stone, W.J.R., M. Eldering, G.-J. van Gemert, K.H.W. Lanke, L. Grignard, M.G. van de Vegte-Bolmer, R. Siebelink-Stoter, W. Graumans, W.F.G. Roeffen, C.J. Drakeley, R.W. Sauerwein, T. Bousema, The relevance and applicability of oocyst prevalence as a read-out for mosquito feeding assays, *Sci. Rep.* 3 (2013) 3418.

[3] Rosenthal, P.J. Antimalarial drug discovery: old and new approaches, *J. Exp. Biol.* 206 (2003) 3735–3744.

[4] Santos, G., N. V. Torres, New Targets for Drug Discovery against Malaria, *PLoS One.* 8 (2013) e59968.

[5] Gnanadesigan, M., V. Nandagopalan, G. Kapildev, M. Gundappa, Nano Drugs for Curing Malaria, in: *Appl. Target. Nano Drugs Deliv. Syst.,* Elsevier, 2019: pp. 451–467.

[6] Kumar, A., K. Srivastava, S. Raja Kumar, M.I. Siddiqi, S.K. Puri, J.K. Sexana, P.M.S. Chauhan, 4-Anilinoquinoline triazines: A novel class of hybrid antimalarial agents, *Eur. J. Med. Chem.* 46 (2011) 676–690.

[7] O'Neill, P.M., V.E. Barton, S.A. Ward, The Molecular Mechanism of Action of Artemisinin—The Debate Continues, *Molecules.* 15 (2010) 1705–1721.

[8] Giobbia, M., E. Tonon, A. Zanatta, L. Cesaris, Z. Bisoffi, A. Vaglia, Late recrudescence of Plasmodium falciparum malaria in a pregnant woman: a case report, *Int. J. Infect. Dis.* 9 (2005) 234–235.

[9] Baird, J.K., S.L. Hoffman, Primaquine Therapy for Malaria, *Clin. Infect. Dis.* 39 (2004) 1336–1345.

[10] Ashley, E.A., J. Recht, N.J. White, Primaquine: the risks and the benefits, *Malar. J.* 13 (2014) 418.

[11] Wellems, T.E., C. V. Plowe, Chloroquine-Resistant Malaria, *J. Infect. Dis.* 184 (2001) 770–776.

[12] Paloque, L., A.P. Ramadani, O. Mercereau-Puijalon, J.-M. Augereau, F. Benoit-Vical, Plasmodium falciparum: multifaceted resistance to artemisinins, *Malar. J.* 15 (2016) 149.

[13] Watts, G. Nobel awarded to discoverers of ivermectin and artemisinin, *BMJ.* (2015) h5352.

[14] Õmura, S., A. Crump, The life and times of ivermectin — a success story, *Nat. Rev. Microbiol.* 2 (2004) 984–989.

[15] Wilson, M.L. Avermectins in arthropod vector management — prospects and pitfalls, *Parasitol. Today.* 9 (1993) 83–87.

[16] S. Pampiglione, G. Majori, G. Petrangeli, R. Romi, Avermectins, MK-933 and MK-936, for mosquito control, *Trans. R. Soc. Trop. Med. Hyg.* 79 (1985) 797–799.

[17] R.W. Burg, B.M. Miller, E.E. Baker, J. Birnbaum, S.A. Currie, R. Hartman, Y.-L. Kong, R.L. Monaghan, G. Olson, I. Putter, J.B. Tunac, H. Wallick, E.O. Stapley, R. Oiwa, S. Ōmura, Avermectins, New Family of Potent Anthelmintic Agents: Producing Organism and Fermentation, *Antimicrob. Agents Chemother. 1*5 (1979) 361–367.

[18] Miller, T.W., L. Chaiet, D.J. Cole, L.J. Cole, J.E. Flor, R.T. Goegelman, V.P. Gullo, H. Joshua, A.J. Kempf, W.R. Krellwitz, R.L. Monaghan, R.E. Ormond, K.E. Wilson, G. Albers-Schönberg, I. Putter, Avermectins, New Family of Potent Anthelmintic Agents: Isolation and Chromatographic Properties, *Antimicrob. Agents Chemother.* 15 (1979) 368–371.

[19] Tavul, L., M. Laman, C. Howard, B. Kotty, A. Samuel, C. Bjerum, K. O'Brian, S. Kumai, M. Amuga, L. Lorry, Z. Kerry, M. Kualawi, S. Karl, L. Makita, L.N. John, S. Bieb, J. Wangi, G.J. Weil, C.W. Goss, D.J. Tisch, W. Pomat, C.L. King, L.J. Robinson, Safety and efficacy of mass drug administration with a single-dose triple-drug regimen of albendazole + diethylcarbamazine + ivermectin for lymphatic filariasis in Papua New Guinea: An open-label, cluster-randomised trial, *PLoS Negl. Trop. Dis.* 16 (2022) e0010096.

[20] Centers for Disease Control and Prevention (CDC), Progress toward elimination of onchocerciasis in the Americas - 1993-2012., MMWR. *Morb. Mortal. Wkly. Rep.* 62 (2013)

[21] Rebollo, M.P., M.J. Bockarie, Toward the elimination of lymphatic filariasis by 2020: treatment update and impact assessment for the endgame, *Expert Rev. Anti. Infect. Ther.* 11 (2013) 723–731.

[22] Richards, F.O. Upon entering an age of global ivermectin-based integrated mass drug administration for neglected tropical diseases and malaria, *Malar. J.* 16 (2017) 168.

[23] Okonkwo, P.O., J.E. Ogbuokiri, E. Ofoegbu, U. Klotz, Protein binding and ivermectin estimations in patients with onchocerciasis, *Clin. Pharmacol. Ther.* 53 (1993) 426–430.

[24] Chaccour, C., F. Hammann, N.R. Rabinovich, Ivermectin to reduce malaria transmission I. Pharmacokinetic and pharmacodynamic considerations regarding efficacy and safety, *Malar. J.* 16 (2017) 161.

[25] Chaccour, C., N.R. Rabinovich, Ivermectin to reduce malaria transmission II. Considerations regarding clinical development pathway, *Malar. J.* 16 (2017) 166.

[26] Tesh, R.B., H. Guzman, Mortality and Infertility in Adult Mosquitoes After the Ingestion of Blood Containing Ivermectin, *Am. J. Trop. Med. Hyg.* 43 (1990) 229–233.

[27] Chaccour, C., J. Lines, C.J.M. Whitty, Effect of Ivermectin on Anopheles gambiae Mosquitoes Fed on Humans: The Potential of Oral Insecticides in Malaria Control, *J. Infect. Dis.* 202 (2010) 113–116.

[28] Panchal, M., K. Rawat, G. Kumar, K.M. Kibria, S. Singh, M. Kalamuddin, A. Mohmmed, P. Malhotra, R. Tuteja, Plasmodium falciparum signal recognition particle components and anti-parasitic effect of ivermectin in blocking nucleo-cytoplasmic shuttling of SRP, *Cell Death Dis.* 5 (2014) e994–e994.

[29] Miller, T.W., L. Chaiet, D.J. Cole, L.J. Cole, J.E. Flor, R.T. Goegelman, V.P. Gullo, H. Joshua, A.J. Kempf, W.R. Krellwitz, R.L. Monaghan, R.E. Ormond, K.E. Wilson, G. Albers-Schönberg, I. Putter, Avermectins, New Family of Potent Anthelmintic Agents: Isolation and Chromatographic Properties, *Antimicrob. Agents Chemother.* 15 (1979) 368–371.

[30] Egerton, J.R., D.A. Ostlind, L.S. Blair, C.H. Eary, D. Suhayda, S. Cifelli, R.F. Riek, W.C. Campbell, Avermectins, New Family of Potent Anthelmintic Agents: Efficacy of the B 1a Component, Antimicrob. *Agents Chemother.* 15 (1979) 372–378.

[31] Campbell, W.C. History of Avermectin and Ivermectin, with Notes on the History of Other Macrocyclic Lactone Antiparasitic Agents, *Curr. Pharm. Biotechnol.* 13 (2012) 853–865.

[32] Zgolicz, P.D., M.I. Cabrera, R.J. Grau, Insight into phosphine effects on the homogeneous hydrogenation of avermectins to ivermectin catalyzed by in-situ formed rhodium complexes, *React. Kinet. Catal. Lett.* 93 (2008) 165–173.

[33] Õmura, S. Ivermectin: 25 years and still going strong, *Int. J. Antimicrob. Agents.* 31 (2008) 91–98.

[34] Billingsley, P., Binka, F. et al. A Roadmap for the Development of Ivermectin as a Complementary Malaria Vector Control Tool, *Am. J. Trop. Med. Hyg.* 102 (2020) 3–24.

[35] Cupp, E.W., M. Sauerbrey, F. Richards, Elimination of human onchocerciasis: History of progress and current feasibility using ivermectin (Mectizan®) monotherapy, *Acta Trop. 1*20 (2011) S100–S108.

[36] Ali, R., S. Salam, *Chemistry and biological activities of ivermectin,* Scrivener Publisher (Wiley), (2023), ISBN: 9781394166541.

[37] Alvi, S., M.K. Hussain, R. Ali, *Chapter-02, Historical background, and synthetic approaches to ivermectin (IVM) & its homologues,* (2023), DOI: 10.1002/9781394168033.ch2.

[38] Crump, A., S. Omura, Ivermectin, "Wonder drug" from Japan: the human use perspective, *Proc. Japan Acad. Ser. B.* 87 (2011) 13–28.

[39] Meyers, J.I., M. Gray, W. Kuklinski, L.B. Johnson, C.D. Snow, W.C. Black, K.M. Partin, B.D. Foy, Characterization of the target of ivermectin, the glutamate-gated chloride channel, from Anopheles gambiae, *J. Exp. Biol.* 218 (2015) 1478–1486.

[40] Ménez, C., J.-F. Sutra, R. Prichard, A. Lespine, Relative Neurotoxicity of Ivermectin and Moxidectin in Mdr1ab (−/−) Mice and Effects on Mammalian GABA(A) Channel Activity, *PLoS Negl. Trop. Dis.* 6 (2012) e1883.

[41] Fritz, M.L., P.Y. Siegert, E.D. Walker, M.N. Bayoh, J.R. Vulule, J.R. Miller, Toxicity of bloodmeals from ivermectin-treated cattle to Anopheles gambiae s.l., *Ann. Trop. Med. Parasitol.* 103 (2009) 539–547.

[42] Nasveld, P., B. Russell, B. Kotecka, K. Rieckmann, Lack of in vitro effect of ivermectin on Plasmodium falciparum., *Southeast Asian J. Trop. Med. Public Health.* 34 (2003) 552–3.

[43] Larivière, M., B. Beauvais, M. Aziz, J.F. Garin, J. Abeloos, F. Derouin, M. Bamba, C. Bossebceuf, M. Ferly-Therizol, C. Sarfati, [A study in the Ivory Coast (1985-1987) of the efficacy and tolerance of ivermectin (Mectizan) in human onchocerciasis. I. A comparative double-blind study of 220 patients with onchocerciasis treated with a single oral dose of 100, 150 or 200 mcg/kg]., *Bull. Soc. Pathol. Exot. Filiales.* 82 (1989) 35–47.

[44] Kobylinski, K.C., B.D. Foy, J.H. Richardson, Ivermectin inhibits the sporogony of Plasmodium falciparum in Anopheles gambiae, *Malar. J.* 11 (2012) 381.

[45] Ouedraogo, A.L., G.J.H. Bastiaens, A.B. Tiono, W.M. Guelbeogo, K.C. Kobylinski, A. Ouedraogo, A. Barry, E.C. Bougouma, I. Nebie, M.S. Ouattara, K.H.W. Lanke, L. Fleckenstein, R.W. Sauerwein, H.C. Slater, T.S. Churcher, S.B. Sirima, C. Drakeley, T. Bousema, Efficacy and Safety of the Mosquitocidal Drug Ivermectin to Prevent Malaria Transmission After Treatment: A Double-Blind, Randomized, Clinical Trial, *Clin. Infect. Dis.* 60 (2015) 357–365.

[46] Pooda, H.S., J.-B. Rayaisse, D.F. de S. Hien, T. Lefèvre, S.R. Yerbanga, Z. Bengaly, R.K. Dabiré, A.M.G. Belem, I. Sidibé, P. Solano, K. Mouline, Administration of ivermectin to peridomestic cattle: a promising approach to target the residual transmission of human malaria, *Malar. J.* 14 (2015) 496.

[47] Alout, H., B.J. Krajacich, J.I. Meyers, N.D. Grubaugh, D.E. Brackney, K.C. Kobylinski, J.W. Diclaro, F.K. Bolay, L.S. Fakoli, A. Diabaté, R.K. Dabiré, R.W. Bougma, B.D. Foy, Evaluation of ivermectin mass drug administration for malaria transmission control across different West African environments, *Malar. J.* 13 (2014) 417.

[48] Kiszewski, A., A. Mellinger, A. Spielman, P. Malaney, S.E. Sachs, J. Sachs, A global index representing the stability of malaria transmission., *Am. J. Trop. Med. Hyg.* 70 (2004) 486–98.

[49] Tesh, R.B., H. Guzman, Mortality and Infertility in Adult Mosquitoes After the Ingestion of Blood Containing Ivermectin, *Am. J. Trop. Med. Hyg.* 43 (1990) 229–233.

[50] Sylla, M., K.C. Kobylinski, M. Gray, P.L. Chapman, M.D. Sarr, J.L. Rasgon, B.D. Foy, Mass drug administration of ivermectin in south-eastern Senegal reduces the survivorship of wild-caught, blood fed malaria vectors, *Malar. J.* 9 (2010) 365.

[51] Foley, D.H., J.H. Bryan, G.W. Lawrence, The potential of ivermectin to control the malaria vector Anopheles farauti, *Trans. R. Soc. Trop. Med. Hyg.* 94 (2000) 625–628.

[52] Kositz, C., J. Talina, J. Diau, R. Asugeni, C. Whitehorn, D. Mabey, C. Chaccour, M. Marks, Incidental mosquitocidal effect of an ivermectin mass drug administration on Anopheles farauti conducted for scabies control in the Solomon Islands, *Trans. R. Soc. Trop. Med. Hyg.* 111 (2017) 97–101.

[53] Bockarie, M.J., J.L.K. Hii, N.D.E. Alexander, F. Bockarie, H. Dagoro, J.W. Kazura, M.P. Alpers, Mass treatment with ivermectin for filariasis control in Papua New Guinea: impact on mosquito survival, *Med. Vet. Entomol.* 13 (1999) 120–123.

[54] Foy, B.D., K.C. Kobylinski, I.M. da Silva, J.L. Rasgon, M. Sylla, Endectocides for malaria control, *Trends Parasitol.* 27 (2011) 423–428.

[55] Fritz, M.L., P.Y. Siegert, E.D. Walker, M.N. Bayoh, J.R. Vulule, J.R. Miller, Toxicity of bloodmeals from ivermectin-treated cattle to Anopheles gambiae s.l., *Ann. Trop. Med. Parasitol.* 103 (2009) 539–547.

[56] Singh, L., K. Singh, Ivermectin: A Promising Therapeutic for Fighting Malaria. Current Status and Perspective, *J. Med. Chem.* 64 (2021) 9711–9731.

[57] White, N.J. Determinants of relapse periodicity in Plasmodium vivax malaria, *Malar. J.* 10 (2011) 297.

[58] Watson, J., W.R.J. Taylor, G. Bancone, C.S. Chu, P. Jittamala, N.J. White, Implications of current therapeutic restrictions for primaquine and tafenoquine in the radical cure of vivax malaria, *PLoS Negl. Trop. Dis.* 12 (2018) e0006440.

[59] Baird, J.K., K.E. Battle, R.E. Howes, Primaquine ineligibility in anti-relapse therapy of Plasmodium vivax malaria: the problem of G6PD deficiency and cytochrome P-450 2D6 polymorphisms, *Malar. J.* 17 (2018) 42.

[60] Ferreira, M.U., T.C. de Oliveira, Challenges for Plasmodium vivax malaria elimination in the genomics era, *Pathog. Glob. Health.* 109 (2015) 89–90.

[61] da Silva, A.N., C.C. Santos, R.N. Lacerda, R.L. Machado, M.M. Póvoa, Susceptibility of Anopheles aquasalis and An. darlingi to Plasmodium vivax VK210 and VK247, *Mem. Inst. Oswaldo Cruz.* 101 (2006) 547–550.

[62] Rios-Velásquez, C.M., K.M. Martins-Campos, R.C. Simões, T. Izzo, E. V dos Santos, F.A. Pessoa, J.B. Lima, W.M. Monteiro, N.F. Secundino, M.V. Lacerda, W.P. Tadei, P.F. Pimenta, Experimental Plasmodium vivax infection of key Anopheles species from the Brazilian Amazon, *Malar. J.* 12 (2013) 460.

[63] Kobylinski, K.C., K.S. Escobedo-Vargas, V.M. López-Sifuentes, S. Durand, E.S. Smith, G.C. Baldeviano, R. V. Gerbasi, S.-B. Ballard, C.A. Stoops, G.M. Vásquez, Ivermectin susceptibility, sporontocidal effect, and inhibition of time to re-feed in the Amazonian malaria vector Anopheles darlingi, *Malar. J.* 16 (2017) 474.

[64] Eastman, R.T., D.A. Fidock, Artemisinin-based combination therapies: a vital tool in efforts to eliminate malaria, *Nat. Rev. Microbiol.* 7 (2009) 864–874.

[65] Sawa, P., S.A. Shekalaghe, C.J. Drakeley, C.J. Sutherland, C.K. Mweresa, A.Y. Baidjoe, A. Manjurano, R.A. Kavishe, K.B. Beshir, R.U. Yussuf, S.A. Omar, C.C. Hermsen, L. Okell, H.D.F.H. Schallig, R.W. Sauerwein, R.L. Hallett, T. Bousema, Malaria Transmission After Artemether-Lumefantrine and Dihydroartemisinin-Piperaquine: A Randomized Trial, *J. Infect. Dis.* 207 (2013) 1637–1645.

[66] Noedl, H., Y. Se, K. Schaecher, B.L. Smith, D. Socheat, M.M. Fukuda, Evidence of Artemisinin-Resistant Malaria in Western Cambodia, N. *Engl. J. Med.* 359 (2008) 2619–2620.

[67] Ouedraogo, A.L., G.J.H. Bastiaens, A.B. Tiono, W.M. Guelbeogo, K.C. Kobylinski, A. Ouedraogo, A. Barry, E.C. Bougouma, I. Nebie, M.S. Ouattara, K.H.W. Lanke, L. Fleckenstein, R.W. Sauerwein, H.C. Slater, T.S. Churcher, S.B. Sirima, C. Drakeley, T. Bousema, Efficacy and Safety of the Mosquitocidal Drug

Ivermectin to Prevent Malaria Transmission After Treatment: A Double-Blind, Randomized, Clinical Trial, *Clin. Infect. Dis.* 60 (2015) 357–365.

[68] Baraka, O.Z., B.M. Mahmoud, C.K. Marschke, T.G. Geary, M.M.A. Homeida, J.F. Williams, Ivermectin distribution in the plasma and tissues of patients infected with Onchocerca volvulus, *Eur. J. Clin. Pharmacol.* 50 (1996) 407–410.

[69] Sampaio, V.S., T.P. Beltrán, K.C. Kobylinski, G.C. Melo, J.B.P. Lima, S.G.M. Silva, Í.C. Rodriguez, H. Silveira, M.G.V.B. Guerra, Q. Bassat, P.F.P. Pimenta, M.V.G. Lacerda, W.M. Monteiro, Filling gaps on ivermectin knowledge: effects on the survival and reproduction of Anopheles aquasalis, a Latin American malaria vector, *Malar. J.* 15 (2016) 491.

[70] Lyimo, I.N., S.T. Kessy, K.F. Mbina, A.A. Daraja, L.L. Mnyone, Ivermectin-treated cattle reduces blood digestion, egg production and survival of a free-living population of Anopheles arabiensis under semi-field condition in south-eastern Tanzania, *Malar. J.* 16 (2017) 239.

[71] Mekuriaw, W., M. Balkew, L.A. Messenger, D. Yewhalaw, A. Woyessa, F. Massebo, The effect of ivermectin® on fertility, fecundity and mortality of Anopheles arabiensis fed on treated men in Ethiopia, *Malar. J.* 18 (2019) 357.

[72] Pasay, C.J., L. Yakob, H.R. Meredith, R. Stewart, P.C. Mills, M.H. Dekkers, O. Ong, S. Llewellyn, R.L.E. Hugo, J.S. McCarthy, G.J. Devine, Treatment of pigs with endectocides as a complementary tool for combating malaria transmission by Anopheles farauti (s.s.) in Papua New Guinea, *Parasit. Vectors.* 12 (2019) 124.

[73] Dreyer, S.M., D. Leiva, M. Magaña, M. Pott, J. Kay, A. Cruz, N.L. Achee, J.P. Grieco, J.A. Vaughan, Fipronil and ivermectin treatment of cattle reduced the survival and ovarian development of field-collected Anopheles albimanus in a pilot trial conducted in northern Belize, *Malar. J.* 18 (2019) 296.

[74] González Canga, A., A.M. Sahagún Prieto, M.J. Diez Liébana, N. Fernández Martínez, M. Sierra Vega, J.J. García Vieitez, The Pharmacokinetics and Interactions of Ivermectin in Humans—A Mini-review, *AAPS J.* 10 (2008) 42–46.

[75] Chaccour, C., Á.I. Barrio, A.G.G. Royo, D.M. Urbistondo, H. Slater, F. Hammann, J.L. Del Pozo, Screening for an ivermectin slow-release formulation suitable for malaria vector control, *Malar. J.* 14 (2015) 102.

[76] Chaccour, C., G. Abizanda, Á. Irigoyen, J.L. Del Pozo, Pilot Study of a Slow-Release Ivermectin Formulation for Malaria Control in a Pig Model, *Antimicrob. Agents Chemother.* 61 (2017).

[77] Bellinger, A.M., M. Jafari, T.M. Grant, S. Zhang, H.C. Slater, E.A. Wenger, S. Mo, Y.-A.L. Lee, H. Mazdiyasni, L. Kogan, R. Barman, C. Cleveland, L. Booth, T. Bensel, D. Minahan, H.M. Hurowitz, T. Tai, J. Daily, B. Nikolic, L. Wood, P.A. Eckhoff, R. Langer, G. Traverso, Oral, ultra–long-lasting drug delivery: Application toward malaria elimination goals, *Sci. Transl. Med.* 8 (2016).

[78] Chaccour, C., N.R. Rabinovich, Ivermectin to reduce malaria transmission III. Considerations regarding regulatory and policy pathways, *Malar. J.* 16 (2017) 162.

[79] Kobylinski, K.C., P. Jittamala, B. Hanboonkunupakarn, S. Pukrittayakamee, K. Pantuwatana, S. Phasomkusolsil, S.A. Davidson, M. Winterberg, R.M. Hoglund, M. Mukaka, R.W. Pluijm, A. Dondorp, N.P.J. Day, N.J. White, J. Tarning, Safety, Pharmacokinetics, and Mosquito-Lethal Effects of Ivermectin in Combination With

Dihydroartemisinin-Piperaquine and Primaquine in Healthy Adult Thai Subjects, *Clin. Pharmacol. Ther.* 107 (2020) 1221–1230.

[80] Walter, P., G. Blobel, Signal recognition particle contains a 7S RNA essential for protein translocation across the endoplasmic reticulum, *Nature.* 299 (1982) 691–698.

[81] Mendes, A.M., I.S. Albuquerque, M. Machado, J. Pissarra, P. Meireles, M. Prudêncio, Inhibition of Plasmodium Liver Infection by Ivermectin, Antimicrob. *Agents Chemother.* 61 (2017).

[82] Vanachayangkul, P., R. Im-erbsin, A. Tungtaeng, C. Kodchakorn, A. Roth, J. Adams, C. Chaisatit, P. Saingam, R.J. Sciotti, G.A. Reichard, C.K. Nolan, B.S. Pybus, C.C. Black, L.A. Lugo-Roman, M.D. Wegner, P.L. Smith, M. Wojnarski, B.A. Vesely, K.C. Kobylinski, Safety, Pharmacokinetics, and Activity of High-Dose Ivermectin and Chloroquine against the Liver Stage of Plasmodium cynomolgi Infection in Rhesus Macaques, Antimicrob. *Agents Chemother.* 64 (2020).

[83] Azevedo, R., A.M. Mendes, M. Prudêncio, Inhibition of Plasmodium sporogonic stages by ivermectin and other avermectins, *Parasit. Vectors.* 12 (2019) 549.

[84] Lalremruata, A., T.T. Nguyen, M.B.B. McCall, G. Mombo-Ngoma, S.T. Agnandji, A.A. Adegnika, B. Lell, M. Ramharter, S.L. Hoffman, P.G. Kremsner, B. Mordmüller, Recombinase Polymerase Amplification and Lateral Flow Assay for Ultrasensitive Detection of Low-Density Plasmodium falciparum Infection from Controlled Human Malaria Infection Studies and Naturally Acquired Infections, *J. Clin. Microbiol.* 58 (2020).

[85] Cordon-Cardo, C., J.P. O'Brien, D. Casals, L. Rittman-Grauer, J.L. Biedler, M.R. Melamed, J.R. Bertino, Multidrug-resistance gene (P-glycoprotein) is expressed by endothelial cells at blood-brain barrier sites., *Proc. Natl. Acad. Sci.* 86 (1989) 695–698.

[86] Chung, K., C.-C. Yang, M.-L. Wu, J.-F. Deng, W.-J. Tsai, Agricultural Avermectins: An Uncommon But Potentially Fatal Cause of Pesticide Poisoning, *Ann. Emerg. Med.* 34 (1999) 51–57.

[87] Nicolas, P., M.F. Maia, Q. Bassat, K.C. Kobylinski, W. Monteiro, N.R. Rabinovich, C. Menéndez, A. Bardají, C. Chaccour, Safety of oral ivermectin during pregnancy: a systematic review and meta-analysis, *Lancet Glob. Heal.* 8 (2020) e92–e100.

[88] Foy, B.D., H. Alout, J.A. Seaman, S. Rao, T. Magalhaes, M. Wade, S. Parikh, D.D. Soma, A.B. Sagna, F. Fournet, H.C. Slater, R. Bougma, F. Drabo, A. Diabaté, A.G. V Coulidiaty, N. Rouamba, R.K. Dabiré, Efficacy and risk of harms of repeat ivermectin mass drug administrations for control of malaria (RIMDAMAL): a cluster-randomised trial, *Lancet.* 393 (2019) 1517–1526.

[89] Awadzi, K., S.K. Attah, E.T. Addy, N.O. Opoku, B.T. Quartey, The effects of high-dose ivermectin regimens on Onchocerca volvulus in onchocerciasis patients, *Trans. R. Soc. Trop. Med. Hyg.* 93 (1999) 189–194.

[90] Makenga, J.-C., Bof, D. Muteba, P. Mansiangi, F. Ilunga-Ilunga, Y. Coppieters, Analysis of severe adverse effects following community-based ivermectin treatment in the Democratic Republic of Congo, BMC Pharmacol. *Toxicol.* 20 (2019) 49.

[91] Bennett, J.L., J.F. Williams, V. Dave, Pharmacology of ivermectin, *Parasitol. Today.* 4 (1988) 226–228.

[92] Mackenzie, C.D., M. Boussinesq, T.G. Geary, R.K. Prichard, J. Kamgno, C. Bourguinat, Analysis of the mdr-1 Gene in Patients Co-Infected with Onchocerca volvulus and Loa loa Who Experienced a Post-Ivermectin Serious Adverse Event, *Am. J. Trop. Med. Hyg.* 83 (2010) 28–32.

[93] Muñoz, J., M.R. Ballester, R.M. Antonijoan, I. Gich, M. Rodríguez, E. Colli, S. Gold, A.J. Krolewiecki, Safety and pharmacokinetic profile of fixed-dose ivermectin with an innovative 18mg tablet in healthy adult volunteers, *PLoS Negl. Trop. Dis.* 12 (2018) e0006020.

[94] Osei-Atweneboana, M.Y., K. Awadzi, S.K. Attah, D.A. Boakye, J.O. Gyapong, R.K. Prichard, Phenotypic Evidence of Emerging Ivermectin Resistance in Onchocerca volvulus, *PLoS Negl. Trop. Dis.* 5 (2011) e998.

Chapter 5

Plant Kunitz-Type Protease Inhibitors: A Review

Rodrigo Michelini de Oliveira Thomasi
José Antônio da Silva
and Ana Hortência Fonsêca Castro*
Universidade Federal de São João del-Rei, *Campus* Centro-Oeste, Divinópolis, Minas Gerais, Brazil

Abstract

Proteases are essential for maintaining homeostasis and controlling physiological processes but have also been linked to pathologies related to the proteolytic systems. To regulate protease activity, animals, plants, and microorganisms produce protease inhibitors (PIs). Due to their inhibitory activities, PIs are involved in immunity, inflammation, resistance to bacterial and fungal infections, blood clotting and fibrinolysis, and tumor immunity, among other activities. For instance, plants produce a varied group known as proteinaceous plant protease inhibitors (PPPIs), which are crucial for preserving physiological homeostasis and supporting natural defense mechanisms. Peptides (≤ 15 kDa), such as Bowman–Birk inhibitors, and small proteins (≥ 15 kDa), such as Kunitz-type protease inhibitors (KTPIs), are examples of PPPI groups that are widely distributed in plants. In this context, legumes are known to contain higher concentrations of KTPIs, especially in seeds. This review highlights the variety of KTPIs, their structures, functions, and biological activities, *in vivo* and *in vitro* production, purification, and characterization methods, and applications in biotechnology.

* Corresponding Author's Email: acastro@ufsj.edu.br

In: Advances in Biology. Volume 9
Editor: Charles D. Grant
ISBN: 979-8-89530-574-4

Keywords: protease inhibitor, Kunitz, biological activities, purification, characterization, biotechnology

Introduction

Proteases are present in all organisms and play a biocatalytic role in protein metabolism, catalyzing highly specific reactions for proteolytic processing. Given the importance of proteins in maintaining homeostasis, proteases are critical in regulating physiological functions (Wang et al., 2024). Several pathological conditions, including cancer, neurological disorders, and inflammatory and cardiovascular diseases, have been associated with changes in proteolytic systems (López-Otín, 2008).

Nature has developed several methods to regulate proteolysis, such as zymogen activation, protease degradation, and protease inhibition by macromolecular inhibitors, called protease inhibitors (PIs) (Farady and Craik, 2010). For instance, plants protect themselves against other species by developing physical and chemical defenses, which involve the production of PIs and non-protein inhibitors that target proteases (Zhu-Salzman and Zeng, 2015).

PIs are divided into approximately 82 families based on sequence homology, and these families have been further divided into 39 clans, according to MEROPS classification (Rawlings et al., 2018). In plants, PIs account for approximately 10% of the total protein content in seeds, tubers, and other tissues, providing the carbon, nitrogen, and sulfur sources required during germination. However, the incidence of these proteins in the aerial parts of plants has also been extensively documented as a result of various stimuli (Cotabarren et al., 2020).

KTPIs belong to the I3 family of peptidase inhibitors and can be found in plants, animals, and microorganisms (Ranasinghe and McManus, 2013). In plants, KTPIs are abundant in the Fabaceae (Leguminosae) family, comprising the Mimosoideae, Caesalpinioideae, and Faboideae/Papilionoideae subfamilies (Oliva et al., 2010).

Distinguished by their stable structure and the presence of a unique domain (Kunitz domain), KTPIs can efficiently inhibit a wide spectrum of proteases, including those from insects and fungi. This adaptability has led to their evolutionary conservation and frequent occurrence in diverse biological systems (Farady and Craik, 2010; Polya, 2003; Ranasinghe and McManus, 2013).

Thus, KTPIs have garnered increasing attention due to their potential applications in drug development and biological activities. Their inhibitory activities have been particularly explored in targeting diseases linked to aberrant protease activity, such as cancer (de Souza et al., 2016; Ranasinghe et al., 2018; Ranasinghe and McManus, 2013) and inflammation (de Magalhães et al., 2018; Machado et al., 2013; Theodoro-Júnior et al., 2017).

There is a trend in the food industry to promote natural alternatives for food preservation. Food science and technology are becoming increasingly interested in natural antibacterial compounds, such as KTPIs, as an alternative form of food preservation because microbes are displaying resistances and becoming more tolerant to food processing and conservation technologies (Cotabarren et al., 2020; Novais et al., 2022)

This burgeoning interest underscores the promise of Kunitz-type inhibitors as versatile tools in both basic research, biotechnology solutions, and therapies. In this regard, this chapter presents a review of the variety of KTPIs, their structures, functions, and biological activities, *in vivo* and *in vitro* production, purification and characterization methods, and applications in biotechnology.

Method

The documents assessed in this review were indexed in electronic databases, including PubMed, Scopus, and Science Direct. The search descriptors were selected using Medical Subject Headings (MeSH) for PubMed. For the search formula, combinations of Boolean operators OR and AND were used, and the results were filtered by language (English and Portuguese). Duplicate documents and those that did not directly address the topic were excluded. After screening the papers by title, abstract, and keywords, all documents were evaluated. Figure 1 depicts the search method in a flowchart.

Plant Kunitz-Type Protease Inhibitors

The identification of KTPIs was a turning point in the study of molecular biology and biochemistry. In 1945, M. Kunitz introduced a 20-kDa protein that inhibited trypsin in soybean seeds (soybean trypsin inhibitor, STI) and purified the protein crystals (Kong et al., 2023; Kunitz, 1945). Other inhibitors with similar qualities were soon identified from plant, microbe, and animal sources (Oliva et al., 2010; Ranasinghe and McManus, 2013).

They function as storage proteins, protect plants against microbial proteases, and control endogenous proteases in plants. KTPIs are generally composed of about 60–80 amino acids and a cysteine-rich motif, which is essential for their inhibitory activity. The conserved pattern of cysteine residues forms disulfide bonds, providing stabilization of the tertiary protein structure (Bendre et al., 2018; Ranasinghe and McManus, 2013).

In general, KTPIs have a molecular weight of 18–24 kDa and a single polypeptide chain with 2 disulfide bonds. They present a single reactive site, β-trefoil fold, and can inhibit serine, cysteine, and aspartic proteinases but are unable to inhibit metalloproteinases (Bendre et al., 2018; Oliva et al., 2010). Their reactive site is generally made up of arginine and lysine but may contain glutamine, alanine, or valine, and they are usually in the region that binds to the enzyme, forming an enzyme-inhibitor complex in a 1:1 stoichiometry (Bonturi et al., 2022).

In KTPIs, the Kunitz domain contains alpha/beta proteins, with three disulfide bonds stabilizing its irregular secondary structure and three peptide loops that can be altered without significantly destabilizing the scaffold. Consequently, KTPIs can tightly bind and form a stable complex with proteases (Oliva et al., 2011; Wang et al., 2023). Kunitz domain-derived inhibitors have been approved by the Food and Drug Administration (FDA). For instance, hereditary angioedema, a rare autosomal inherited blood condition, can be treated with ecallantide (DX-88), a substitute kallikrein inhibitor, when the plasma C1 kallikrein inhibitor does not function. DX-88, derived from the Kunitz domain associated with lipoproteins, was designed to bind kallikrein with picomolar affinity (Mishra, 2020; Simeon and Chen, 2018; Williams and Baird, 2003)

Kunitz inhibitors can be classified depending on their inhibitory profile: serine-specific, cysteine-specific, serine/cysteine, and serine/cysteine/aspartic protease inhibitors. This categorization can aid in the design of inhibitors for different proteases (Mishra, 2020; Smith et al., 2016).

Biological Activities

The biological role of KTPIs is still unclear, but there is evidence that they act as storage proteins, regulators of endogenous proteases, and assist in defense against pests and pathogens (Bendre et al., 2018). Additionally, they can prevent excessive inflammation, regulate cellular activities, inhibit viral proteases, and have antimicrobial activities (Bonturi et al., 2022; de Magalhães et al., 2018; Ershova et al., 2022; Kong et al., 2023; Machado et al., 2013; Manzato et al., 2023; Ranasinghe and McManus, 2013; Cândido et al., 2011; Theodoro-Júnior et al., 2017). Some studies reporting the antimicrobial activities of plant KPTIs are compiled in Table 1.

A 5.6-kDa KTPI was isolated from potato tubers (*Solanum tuberosum* L cv. Gogu) by extraction of the water-soluble fraction, dialysis, ultrafiltration, and C18 reversed-phase high-performance liquid chromatography. This inhibitor, named potamin-1 (PT-1), strongly inhibited pathogenic microbial strains, including *Candida albicans*, *Rhizoctonia solani*, and *Clavibacter michiganensis* ssp. *michiganensis* (Kim et al., 2005).

A trypsin inhibitor from *S. tuberosum* subsp. *andigenum* var. overa, named oPTI (12 kDa), had an inhibitory effect on the growth of *Acinetobacter baumannii* S-1, *Acinetobacter baumannii* R, *Acinetobacter calcoaceticus* R, *Acinetobacter calcoaceticus* S, *Bacillus stearothermophilus*, *Escherichia coli*, *Pseudomonas aeruginosa*, *Salmonella braenderup*, *Salmonella enteritidis*, *Salmonella typhimurium*, and *Yersinia enterocolitica* strains. The oPTI was obtained through *in vitro* production and recombinant expression, purified using a protein production strain (*Escherichia coli* BL21), and characterized using electrophoresis (SDS-PAGE, 12% v/v) (Cisneros et al., 2020).

Two KTPIs from legume seeds (*Vigna radiata* and *Cajanus platycarpus*) were obtained through mild trichloroacetic acid (TCA) extraction followed by trypsin-affinity chromatography and labeled as VrKI (19.2 kDa) and CpKI (19.4 kDa), respectively. The KTPIs were confirmed by immunodetection and matrix-assisted laser desorption ionization (MALDI-TOF) studies and further characterized for their structural (Circular dichroism and fluorescence spectroscopy) and functional properties (temperature and stability). Both KTPIs demonstrated significant potential in controlling the growth of methicillin-sensitive *Staphylococcus aureus*, a Gram-positive pathogenic bacterium (Gujjarlapudi et al., 2023).

Another study reported a KTPI from *Enterolobium timbouva* seeds named EtTI (20 kDa), extracted (1:10, w/v) with 100 mM sodium phosphate, dialyzed, and lyophilized. The activity against different *Candida* strains was assessed using the broth microdilution technique, and high inhibitory activity against trypsin (Ki = 0.5 nM) and refractory to digestion by pepsin were observed. EtTI was active against *C. albicans*, *C. buinensis*, and *C. tropicalis* and triggered disturbances in the plasma membrane integrity (de Oliveira et al., 2018).

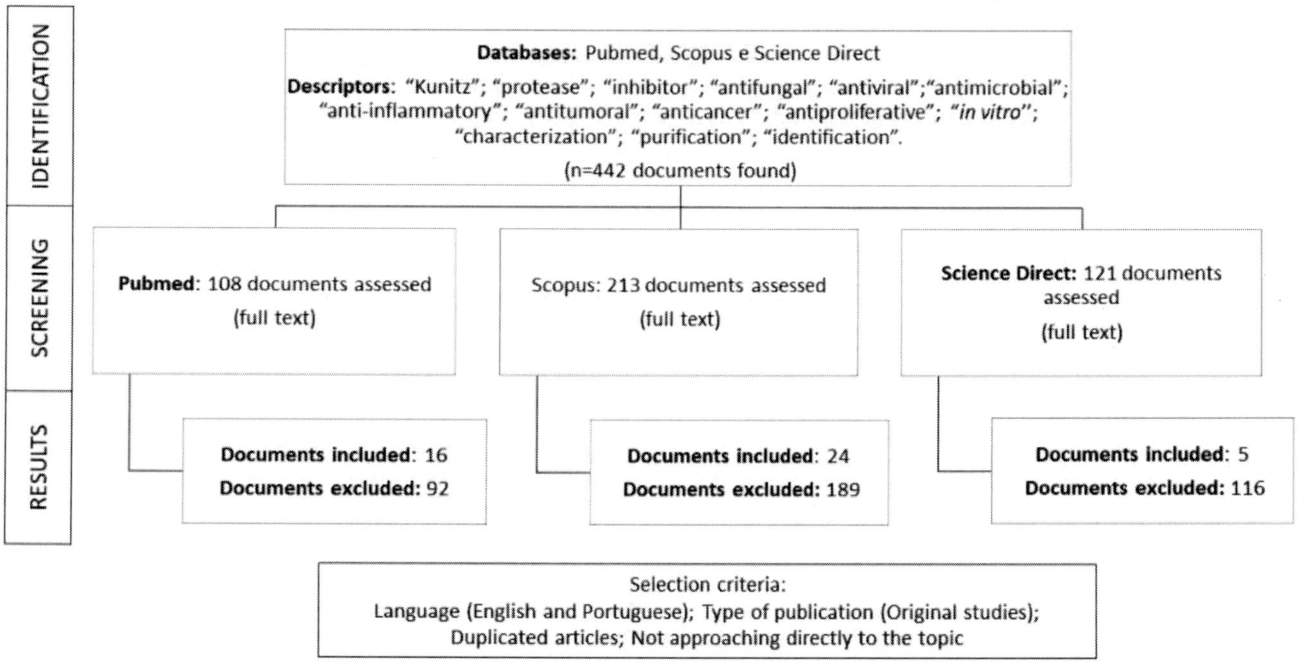

Figure 1. Search flowchart of the steps used, and strategies adopted in this review.

Table 1. Antimicrobial activities of KTPIs

Sources (plant species)	Family	Parts of the plant	Inhibitor name and molecular weight	Inhibitory activity (Targets/ microorganisms)	Obtention and characterization methods	Main assays	Reference
Solanum tuberosum	Solanaceae	Tubers	PT-1 (5,6 kDa)	*Candida albicans* *Rhizoctonia solani* *Clavibacter michiganense*	Extraction of the tubers water-soluble fraction, ultrafiltration, and high performance liquid chromatography (HPLC).	Minimum fungicide concentration (MFC) using broth microdilution technique.	(Kim et al., 2005)
Solanum tuberosum	Solanaceae	Tubers	oPTI (12 kDa)	*Acinetobacter baumannii* *Acinetobacter calcoaceticus* *Bacillus stearothermophilus* *Escherichia coli* *Pseudomonas aeruginosa* *Salmonella braenderup* *Salmonella enteritidis* *Salmonella typhimurium* *Yersinia enterocolitica*	*In vitro* production through recombinant expression, purification though protein production strain (*Escherichia coli* BL21) and characterization through electrophoresis (SDS-PAGE, 12% v/v).	Minimum inhibitory concentration (MIC) assay.	(Cisneros et al., 2020)
Vigna radiata *Cajanus platycarpus*	Fabaceae Fabaceae	Seeds	VrKI (19,2 kDa) CpKI (19,4 kDa)	*Staphylococcus aureus*	Mild trichloroacetic acid (TCA) seed extraction followed by trypsin-affinity chromatography.	Minimum inhibitory concentration (MIC) assay.	(Gujjarlapudi et al., 2023)

Table 1. (Continued)

Sources (plant species)	Family	Parts of the plant	Inhibitor name and molecular weight	Inhibitory activity (Targets/ microorganisms)	Obtention and characterization methods	Main assays	Reference
Enterolobium timbouva	Fabaceae	Seeds	EtTI (20 kDa)	*Candida albicans, Candida buinensis Candida tropicalis*	Extraction with 1:10 w/v sodium phosphate extract, followed by dialysis and lyophilization.	Peptidases inhibitory activity, antifungal growth inhibition, plasma membrane permeabiliztion and ROS induction assays.	(de Oliveira et al., 2018)
Inga vera	Fabaceae	Seeds	IVTI (20 kDa)	*Candida buinensis Escherichia coli*	Extraction with 100 mM potassium phosphate, dialyzed and purificated through diethylaminoethyl (DEAE) sepharose column (2 × 20 cm).	Minimum inhibitory concentration (MIC) and minimum fungicide concentration (MFC) assays.	(Bezerra et al., 2016)
Cassia leiandra	Fabaceae	Seeds	ClTI (19,5 kDa)	*Candida albicans, Candida tropicalis, Candida parapsilosis, Candida krusei*	Water-soluble protein extract (SPE), followed by ion exchange on a DEAE-cellulose column and affinity chromatography on a trypsin-sepharose 4B column affinity matrix.	Serial dilution method, minimum fungicide concentration (MFC) assay. Assessment of biofilm formation and degradation.	(Araújo et al., 2019)
Inga edulis	Fabaceae	Seeds	IETI (19,6 kDa)	*Candida buinensis Candida tropicalis*	Extraction with 100 mM sodium phosphate buffer (1:10, w/v), dialyzed,	Growth inhibition assay, minimum	(Dib et al., 2019)

Sources (plant species)	Family	Parts of the plant	Inhibitor name and molecular weight	Inhibitory activity (Targets/ microorganisms)	Obtention and characterization methods	Main assays	Reference
					lyophilized and purified in two-step column chromatography.	fungicide concentration (MFC).	
Acacia nilotica	Fabaceae	Seeds	AnTI (21 kDa)	*Escherichia coli* *Pseudomonas aeruginosa* *Bacillus subtilis,* *Xanthomonas oryzae,* *Staphylococcus aureus* *Aspergillus niger* *Fusarium oxysporum*	Extraction with Tris buffer (100 mM, pH 7.0), purification by using a combination of anion exchange and gel filtration chromatography. Identification with SDS-PAGE and molecular mass determination.	Microplate assay and inhibition pattern of conidial germination assay.	(Mehmood et al., 2020)
Conyza dioscoridis	Asteraceae	Whole plant	PDInhibitor (25 kDa)	*Escherichia coli* *Bacteroides fragilis* Enterococcus *faecalis* *Salmonella enterica* *Klebsiella pneumonia* *Bacillus cereus* *Staphylococcus epidermidis* Aspergillus *niger* *Botrytis cinerea* *Fusarium solani* *Penicillium digitatum*	0.1 M phosphate buffer (pH 6–7) extraction, Purification with ammonium sulfate precipitation followed by filtration through a Sephadex G-50 column.	Minimum inhibitory concentration (MIC) assay and well diffusion method with Sabouraud dextrose agar.	(Karray et al., 2020)

A KTPI derived from *Ingra vera* seeds (IVTI, 20 kDa) was isolated and described for a variety of biological activities, including insecticidal, antibacterial, and anticancer properties. Its crude extract was obtained by extraction with 100 mM potassium phosphate, dialyzed, and purified using a diethylaminoethyl (DEAE) sepharose column (2 × 20 cm). IVTI inhibited the growth of *Candida buinensis* and *Escherichia coli*, and its activity was assessed using minimum inhibitory concentration (MIC) and minimum fungicide concentration (MFC) assays (Bezerra et al., 2016).

Another KTPI named ClTI (19.5 kDa), from *Cassia leiandra* seeds, was obtained by aqueous extraction using dried powder (10:90, w/v) in 0.15 mol/L NaCl solution and purified through Trypsin Sepharose-CL4B exclusion chromatography and Superdex-G75. ClTI inhibited the growth of *Candida albicans*, *Candida tropicalis*, *Candida parapsilosis*, and *Candida krusei*. ClTI also exhibited antibiofilm activity and non-cytotoxicity to mammalian cells (Araújo et al., 2019).

IETI (19.6 kDa), a KTPI from *Inga edulis*, was extracted (1:10, w/v) with 100 mM sodium phosphate buffer, dialyzed, lyophilized, and purified in a two-step chromatography (Sephadex G-100 column (80 × 3 cm) and trypsin–Sepharose column (4.6 × 80 mm)). IETI showed activity against *C. buinensis* and *C. tropicalis* in a growth inhibition assay (Dib et al., 2019).

Acacia nilotica seeds produce a KTPI named *An*TI (21 kDa). The inhibitor was purified and characterized with LC-MS/MS and SDS-PAGE. *An*TI was purified from seed extract (100 mL Tris buffer, 100 mM, pH 7.0) using a combination of anion exchange and gel filtration chromatography. The antibacterial effect of the purified *An*TI was effective against five human pathogenic bacterial species: *Escherichia coli*, *Pseudomonas aeruginosa*, *Bacillus subtilis*, *Xanthomonas oryzae*, and *Staphylococcus aureus*. In addition, AnTI inhibited *Aspergillus niger* and *Fusarium oxysporum* (Mehmood et al., 2020).

A *Conyza dioscoridis* protease inhibitor was also identified and extracted in 0.1 M phosphate buffer (pH 6–7). PDInhibitor (25 kDa), a KTPI, was purified using ammonium sulfate precipitation followed by filtration through a Sephadex G-50 column and had a molecular weight of 25 kDa. The inhibitor displayed potent antimicrobial activity against Gram-positive and -negative bacteria (Escherichia coli, Bacteroides fragilis, Enterococcus faecalis, Salmonella enterica, Klebsiella pneumonia, Bacillus cereus and Staphylococcus epidermidis) and against fungal strains (Aspergillus niger, Botrytis cinerea, Fusarium solani, and Penicillium digitatum) (Karray et al., 2020).

Table 2. Anti-inflammatory activities of KTPIs

Sources (plant species)	Family	Parts of the plant	Inhibitor name and molecular weight	Anti-inflammatory activity	Obtention and characterization methods	References
Caesalpinia decapetala	Fabaceae	seeds	CdTI (19,6 kDa)	*In vitro* effective decrease in inflammatory mediators (NO, IL-6, TNF-α) and on the activity of elastase in macrophage cell lines (RAW264.7)	Extraction with 0.1 M Tris-HCl buffer pH 7.5. Two-step chromatography technique. Sodium dodecyl sulfate polyacrylamide gel electrophoresis and matrix-assisted laser desorption ionization time of flight.	(Rajegowda and SnehaRani, 2023)
Libidibia ferrea	Fabaceae	Seeds	LfTI (18 kDa)	*In vitro* and *in vivo* ni-inflammatory effects, inhibits neurogenic pain and attenuates inflammatory parameters.	0.1 M Tris-HCl buffer pH 7.5 extraction. Two-step chromatography was used to purify the PIs. Identification with SDS-PAGE electrophoresis and MALDI-TOF.	(Carvalho et al., 2023)
Erythrina velultina	Fabaceae	Seeds	EvTI (19,2 kDa)	Inhibition of leukocyte migration and TNF-α release, stimulating IFN-α and IL-12 synthesis.	1:10 (w/v) in Tris-HCl buffer, pH 7.5 extraction, followed by centrifugation, fractionation and dialysis. Purification with ammonium sulfate (30–60%), fractionation, followed by Trypsin-Sepharose affinity chromatography and reversed-phase HPLC for the identification.	(Machado et al., 2013)
Crataeva tapia	Capparaceae	Bark	CrataBL (21 kDa)	Attenuates changes in lung mechanics, inflammation, extracellular lung remodeling, and oxidative stress responses induced by the administration of elastase	10% (w/v) extraction, fractionated with 30–60% ammonium sulfate, dialyzed and applied to a column of CM-cellulose and exclusion chromatography on Superdex 75 column. Identification by HPLC on a C18 column.	(Oliva et al., 2015)

Table 3. Anticancer activities of KTPIs

Sources (plant species)	Family	Parts of the plant	Inhibitor name and molecular weight	Anticancer/antitumoral/ antiproliferative activity	Obtention and characterization methods	References
Caesalpinia decapetala	Fabaceae	Seeds	CdTI (19,6 kDa)	Antiproliferative against macrophage (RAW264.7) and lung cancer cell lines (A549).	Extraction with 0.1 M Tris-HCl buffer pH 7.5. Two-step chromatography technique. Identification with SDS-PAGE electrophoresis and MALDI-TOF.	(Rajegowda and SnehaRani, 2023)
Inga vera	Fabaceae	Seeds	IVTI (20 kDa)	Prevents the proliferation of CACO-2, a type of cancer cell	Extraction with 100 mM potassium phosphate, dialyzed and purificated through diethylaminoethyl (DEAE) sepharose column (2 × 20 cm).	(Bezerra et al., 2016)
Erythrina velultina	Fabaceae	Seeds	EvCI (19,2 kDa)	Selective anti-proliferative properties against HeLa cells.	0.05 M Tris–HCl buffer, pH 7.5 extraction. Purification by ammonium sulphate fractionation, trypsin– and chymotrypsin–sepharose affinity chromatographies and Resource Q anion-exchange column. Identification with SDS-PAGE and MALDI-TOF.	(Lucena et al., 2022)
Bauhinia purpurea	Fabaceae	Seeds	BPLTI (20 kDa)	Stimulatory effects of BPLTI on tumor cell apoptosis, mitochondrial membrane depolarization, and cytokine production.	Aqueous extraction, purification using ion exchange chromatography on Sp-Sepharose and Mono S and gel filtration on Superdex 75, and identification with SDS-PAGE	(Fang et al., 2012)
Gymnocladus chinensis	Fabaceae	Seeds	GCTI	Inhibition of [methyl-3H] thymidine incorporation by leukemia L1210 and lymphoma MBL2 cells.	Aqueous extraction, purification using ion exchange chromatography (DEAE-cellulose), affinity chromatography on Affi-gel blue gel, ion exchange chromatography on	(Zhu et al., 2011)

Sources (plant species)	Family	Parts of the plant	Inhibitor name and molecular weight	Anticancer/antitumoral/ antiproliferative activity	Obtention and characterization methods	References
					sulfopropyl sepharose (SP sepharose), and gel filtering by FPLC on Superdex 75. Identification using gel electrophoresis (SDS–PAGE)	
Glycine max	Fabaceae	Seeds	KBTI	Immunostimulatory activity, inhibited growth of tumor cell and weakly inhibited proliferation of NPC CNE-2 and HNE-2 cells. It could also cause a slight antiproliferative activity on MCF-7 breast cells and Hep G2 hepatoma cells	Aqueous extraction, purification using Q-Sepharose column Blue-Sepharose column (18 cm×5 cm), and Superdex 75 dialyzed extensively and identification using SDS PAGE 12% polyacrylamide gel.	(Fang et al., 2010)

As demonstrated, the majority of Kunitz type trypsin inhibitors have been found in seeds, due to several reasons related to plant defense and survival, such as defense mechanisms against predators, pathogens and seed longevity (Bonturi et al., 2022; Clemente et al., 2019).

Furthermore, studies reported the anti-inflammatory activities of plant KPTIs, which are compiled in Table 2.

A 19.6-kDa protease inhibitor (PI) isolated from the seeds of *Caesalpinia decapetala* was identified (named CdTI). Defatted seeds were subjected to 0.1 M Tris-HCl buffer pH 7.5 extraction. Two-step chromatography was used to purify the PIs. SDS-PAGE and MALDI-TOF were used to identify CdTI. An *in vitro* test tube assay was performed to determine the anti-inflammatory activity, and CdTI demonstrated antiproliferative activity against macrophage (RAW264.7) and lung cancer cell lines (A549). In addition, a decrease in the production of inflammatory mediators (NO, IL-6, TNF-α) and the activity of elastase was observed in macrophage cell lines (RAW264.7) (Rajegowda and SnehaRani, 2023).

Another study evaluated the anti-inflammatory activity of KTPI purified from *Libidibia ferrea* seeds (LfTI, 18 kDa). It was obtained using extraction with 0.1 M NaCl (10%, wt/mL), followed by dialysis, protein precipitation, purification using two-step chromatography (DEAE sephadex column and trypsin-sepharose 4B resin), and identification using SDS-PAGE. *In vitro* and *in vivo* assays were performed in which male Swiss mice were used for cell harvesting and the *in vivo* assays. The anti-inflammatory activity was analyzed *in vitro* by macrophage cytotoxicity, hydrogen peroxide production, and cell adhesion assays and *in vivo* by leukocyte recruitment, nitric oxide (NO) production, vascular permeability, paw edema, and mast cell degranulation assays. LfTI had ani-inflammatory effects, inhibited neurogenic pain, and attenuated inflammatory parameters (Carvalho et al., 2023).

EvTI (19.2 kDa) is a KTPI obtained and identified using *Erythrina velultina* seed extraction (1:10 (w/v) in Tris-HCl buffer, pH 7.5), centrifugation, fractionation, dialysis, purification using ammonium sulfate (30–60%), fractionation, Trypsin-Sepharose affinity chromatography, and reversed-phase high-performance liquid chromatography. EvTI presented anti-inflammatory properties and prolonged the clotting time in a dose-dependent manner by using *in vitro* and *in vivo* models, inhibiting leukocyte migration and TNF-α release and stimulating IFN-α and IL-12 synthesis (Machado et al., 2013).

Another study assessed the impact of a protein named CrataBL (21 kDa), which was extracted from the bark of the *Crataeva tapia* plant, on lung remodeling and inflammation following elastase-induced pulmonary changes in mice. Using 30–60% ammonium sulfate, 10% (w/v) extracts from *Crataeva tapia* bark were fractionated. The fraction containing the protein was dialyzed and applied to a column of CM-cellulose, and a single peak was submitted to size exclusion chromatography on a Superdex 75 column. This was followed by HPLC on a C18 column to identify the sample. In summary, CrataBL attenuated changes in lung mechanics, inflammation, extracellular lung remodeling, and oxidative stress responses induced by the administration of elastase. It also reduced the volume fraction of isoprostane, collagen, and elastic fibers in the airways and alveolar walls (Oliva et al., 2015).

In addition to anti-inflammatory activities, KTPIs also demonstrate anticancer activities, as shown in Table 3.

As previously mentioned, CdTI (19.6 kDa), isolated from the seeds of *Caesalpinia decapetala* besides the anti-inflammatory activity, also demonstrated antiproliferative activity against macrophage (RAW264.7) and lung cancer cell lines (A549) (Rajegowda and SnehaRani, 2023). In addition to the antimicrobial activities mentioned previously, IVTI (20 kDa) prevents the proliferation of CACO-2, a type of cancer cell (Bezerra et al., 2016).

EvCI (19.2 kDa), also obtained from *Erythrina velultina*, demonstrated selective anti-proliferative properties against HeLa cells *in vitro*. It was obtained through 0.05 M Tris–HCl buffer and pH 7.5 extraction and purified using ammonium sulfate fractionation, trypsin- and chymotrypsin–sepharose affinity chromatography and a resource Q anion-exchange column. It was identified using SDS-PAGE electrophoresis and MALDI-TOF (Lucena et al., 2022).

BPLTI, a 20-kDa KTPI isolated from the seeds of *B. purpurea* L., demonstrated stimulatory effects of BPLTI on tumor cell apoptosis, mitochondrial membrane depolarization, and cytokine production; consequently, it has an inhibitory effect on the proliferation of certain tumor cells. The BPLTI was obtained through aqueous extraction, purified using liquid chromatography processes that included ion exchange chromatography on Sp-Sepharose and Mono S and gel filtration on Superdex 75 and was identified using SDS-PAGE (Fang et al., 2012).

Found in *Gymnocladus chinensis* seeds, a KTPI named GCTI inhibited [methyl-3H] thymidine incorporation by leukemia L1210 and lymphoma MBL2 cells. It was obtained using aqueous extraction. The isolation process consisted of ion exchange chromatography on diethylaminoethyl cellulose (DEAE-cellulose), affinity chromatography on Affi-gel blue gel, ion exchange chromatography on sulfopropyl sepharose (SP sepharose), and gel filtering by

FPLC on Superdex 75. The purified protein was subjected to gel electrophoresis (SDS–PAGE) (Zhu et al., 2011).

Another study used liquid chromatography to purify a KTPI from *Glycine max* (named KBTI) with a molecular weight of 20.10 kDa. This KBTI manifested immunostimulatory activity, inhibited tumor cell growth, and weakly inhibited proliferation of NPC CNE-2 and HNE-2 cells. It can also cause slight antiproliferative activity on MCF-7 breast cells and Hep G2 hepatoma cells. It was obtained using aqueous extraction, applied to a Q-Sepharose column Blue-Sepharose column (18 cm × 5 cm), and Superdex 75). The purified trypsin inhibitor was dialyzed extensively and identified using SDS-PAGE 12% polyacrylamide gel (Fang et al., 2010).

In summary, the extraction solvents more commonly found in this review were aqueous, Tris-HCl buffer and phosphate salts such as potassium phosphate. Many protease inhibitors are water-soluble and aqueous solutions could probably help maintain the integrity and activity of the protease inhibitors (Clemente et al., 2019), and organic solvents or harsh conditions could increase the risk of denaturation (Blanco & Blanco, 2022). The extraction is usually followed by centrifugation, dyalisis and lyophilization, to ensure more purity, concentration, and stability of the proteins for further analysis and applications (Goldring, 2015). As purification process, combinations of different chromatography methods were usually preferred, which is reasonable due to the versatility, specificity, and mild operating conditions of chromatography (Arora et al., 2017; Halan et al., 2019; Kumari et al., 2022). The identification and determination of molecular weight of KTPIs were mainly through SDS-PAGE electrophoresis, which is the most commonly used technology to obtain high resolution analytical separation of mixtures of proteins (Nowakowski et al., 2014).

In Vitro and *In Vivo* IP Production and Applications in Biotechnology

As discussed in the previous sections, plant KTPIs have many biological activities. These properties may provide solutions in food preservation, agriculture, and medicine, among other areas. *In vitro* methods aimed at the production of protease inhibitors have been developed in a sustainable, economically viable and effective manner (Troncoso et al., 2022; Efferth, 2019; González-Rábade et al., 2011).

One of the main techniques that could be applied to KTPI production is *in vitro* plant tissue cultures, which is the aseptic cultivation of cells, tissues, organs, and their components *in vitro* under certain physical and chemical conditions (Troncoso et al., 2022; Efferth, 2019; Thorpe, 2007). Tissues and organs can be cultured for a variety of purposes, including the study of cell behavior (cytology, nutrition, metabolism, morphogenesis, embryogenesis, and pathology), plant modification and improvement, the generation of pathogen-free plants for germplasm storage, clonal propagation, and the production of commercially important metabolites (Kumar and Loh, 2012; Loyola-Vargas and Ochoa-Alejo, 2018) KTPIs and other proteins can be produced through plant tissue culture. For example, callus cultures of *Momordica dioica* have been used to produce serine protease inhibitors *in vitro* (Kumar et al., 2015).

To increase protease volumetric production, *in vitro* tissue cultures can also be performed with genetically engineered plants (Karki et al., 2021). Relevant growing culture parameters include transgenic cell line selection, temperature, pH, oxygen demand, and bioreactor type transferred to the field. The Morishige and Skoog medium (in combination with kinetin and indole butyric acid), for example, is widely used in plant cell cultures (Troncoso et al., 2022; González-Rábade et al., 2011). In addition, callus cultures can be converted to cell suspension cultures on shakers and bioreactors, allowing growth and metabolite production without the influence of environmental factors, seasonal variation microbial diseases, pests, and geographical constraints (David Troncoso et al., 2022; Santos et al., 2016; Winson et al., 2020).

Cell suspension cultures (CSCs), in bioreactors for example, have several benefits considering the sustained biopharmaceutical production over conventional microbial and/or mammalian host systems, including inherent safety, affordable bioprocessing, and the ability to post-translationally modify proteins (Huang and McDonald, 2009). Plant CSCs are a good option for producing, for example, recombinant proteins (such as recombinant KTPIs) since it combines the benefits of microbial and entire plant-based systems. They are cost-effective, simple to scale up, and propagable in aseptic environments (Rani and Vimolmangkang, 2024).

Conclusion

Studies have demonstrated the complexity of protease inhibitors and their biological roles. In this review, we described some of the biological activities of protease inhibitors, including antibacterial, antifungal, inhibition of inflammatory enzymes, and *in vitro* reduction of tumor cell invasion and proliferation. Since plant KTPIs exhibit a wide range of biological actions, these qualities may hold answers for various problems in the fields of agriculture, medicine, and biotechnology. Thus, *in vitro* culture techniques are an alternative to producing protease inhibitors in a sustainable, profitable, and efficient way.

Acknowledgements

The authors are grateful to the Universidade Federal de São João del-Rei (UFSJ), Fundação de Amparo à Pesquisa do Estado de Minas Gerais (FAPEMIG - APQ-00861-22 and APQ-04400-23), and Coordenação de Aperfeiçoamento de Pessoal de Nível Superior (CAPES) for financial support. R.M.O.T is grateful to FAPEMIG for scholarship award. This study was financed in part by the Coordenação de Aperfeiçoamento de Pessoal de Nível Superior – Brasil (CAPES) – Finance Code 001.

Disclaimer

None

References

Araújo, N. M. S., Dias, L. P., Costa, H. P. S., Sousa, D. O. B., Vasconcelos, I. M., de Morais, G. A., and Oliveira, J. T. A. 2019. ClTI, a Kunitz trypsin inhibitor purified from *Cassia leiandra* Benth. seeds, exerts a candidicidal effect on *Candida albicans* by inducing oxidative stress and necrosis. *Biochimica et Biophysica Acta (BBA) - Biomembranes*, *1861*(11), 183032. https://doi.org/10.1016/J.BBAMEM.2019.183032

Arora, S., Saxena, V., and Ayyar, B. V. 2017. Affinity chromatography: A versatile technique for antibody purification. *Methods*, 116, 84–94. https://doi.org/10.1016/J.YMETH.2016.12.010

Bendre, A. D., Ramasamy, S., and Suresh, C. G. 2018. Analysis of Kunitz inhibitors from plants for comprehensive structural and functional insights. *International Journal of Biological Macromolecules*, 113, 933–943. https://doi.org/10.1016/J.IJBIOMAC.2018.02.148

Bezerra, C. S., De Oliveira, C. F. R., Machado, O. L. T., De Mello, G. S. V., Pitta, M. G. R., Rêgo, M. J. B. M., Napoleão, T. H., Paiva, P. M. G., Ribeiro, S. F. F., Gomes, V. M., Silva, O. N., Maria-Neto, S., Franco, O. L., and Macedo, M. L. R. 2016. Exploiting the biological roles of the trypsin inhibitor from *Inga vera* seeds: A multifunctional Kunitz inhibitor. *Process Biochemistry*, *51*(6), 792–803. https://doi.org/10.1016/J.PROCBIO.2016.03.008

Blanco, A., and Blanco, G. 2022. "Proteins". *Medical Biochemistry*, 21–75. https://doi.org/10.1016/B978-0-323-91599-1.00004-3

Bonturi, C. R., Teixeira, A. B. S., Rocha, V. M., Valente, P. F., Oliveira, J. R., Filho, C. M. B., Batista, I. de F. C., and Oliva, M. L. V. 2022. Plant Kunitz Inhibitors and Their Interaction with Proteases: Current and Potential Pharmacological Targets. *International Journal of Molecular Sciences*, 23(9). https://doi.org/10.3390/IJMS23094742

Cândido, E., Pinto, M. F. S., Pelegrini, P. B., Lima, T. B., Silva, O. N., Pogue, R., Grossi-de-Sá, M. F., and Franco, O. L. 2011. Plant storage proteins with antimicrobial activity: novel insights into plant defense mechanisms. *FASEB Journal : Official Publication of the Federation of American Societies for Experimental Biology*, 25(10), 3290–3305. https://doi.org/10.1096/FJ.11-184291

Carvalho, R., Bonfá, I. S.,Muller, J. I. A., Pando, S. C., and Toffoli-Kadri, M. C. 2023. Protease inhibitor from *Libidibia ferrea* seeds attenuates inflammatory and nociceptive responses in mice. *Journal of Ethnopharmacology*, 300, 115694. https://doi.org/10.1016/J.JEP.2022.
115694

Cisneros, J. S., Cotabarren, J., Parisi, M. G., Vasconcelos, M. W., and Obregón, W. D. 2020. Purification and characterization of a novel trypsin inhibitor from *Solanum tuberosum* subsp. *andigenum* var. *overa*: Study of the expression levels and preliminary evaluation of its antimicrobial activity. *International Journal of Biological Macromolecules*, *158*, 1279–1287. https://doi.org/10.1016/J.IJBIOMAC.2020.04.217

Clemente, M., Corigliano, M. G., Pariani, S. A., Sánchez-López, E. F., Sander, V. A., and Ramos-Duarte, V. A. 2019. Plant Serine Protease Inhibitors: Biotechnology Application in Agriculture and Molecular Farming. *International Journal of Molecular Sciences*, *20*(6). https://doi.org/10.3390/IJMS20061345

Cotabarren, J., Lufrano, D., Parisi, M. G., and Obregón, W. D. 2020. Biotechnological, biomedical, and agronomical applications of plant protease inhibitors with high stability: A systematic review. *Plant Science*, 292, 110398. https://doi.org/10.1016/J.PLANTSCI.2019.110398

de Magalhães, M. T. Q., Mambelli, F. S., Santos, B. P. O., Morais, S. B., and Oliveira, S. C. 2018. Serine protease inhibitors containing a Kunitz domain: their role in modulation of host inflammatory responses and parasite survival. *Microbes and Infection*, 20(9–10), 606–609. https://doi.org/10.1016/J.MICINF.2018.01.003

de Oliveira, C. F. R., Oliveira, C. T., Taveira, G. B., Mello, E. O., Gomes, V. M., and Macedo, M. L. R. 2018. Characterization of a Kunitz trypsin inhibitor from *Enterolobium timbouva* with activity against *Candida* species. *International Journal of Biological Macromolecules*, 119, 645–653. https://doi.org/10.1016/J.IJBIOMAC.2018.07.034

de Souza, J. G., Morais, K. L., Anglés-Cano, E., Boufleur, P., de Mello, E. S., Maria, D. A., Origassa, C. S. T., Zampolli, H. de C., Câmara, N. O. S., Berra, C. M., Bosch, R. V., and Chudzinski-Tavassi, A. M. 2016. Promising pharmacological profile of a Kunitz-type inhibitor in murine renal cell carcinoma model. *Oncotarget*, 7(38), 62255–62266. https://doi.org/10.18632/ONCOTARGET.11555

Dib, H. X., de Oliveira, D. G. L., de Oliveira, C. F. R., Taveira, G. B., de Oliveira Mello, E., Verbisk, N. V., Chang, M. R., Corrêa-Junior, D., Gomes, V. M., and Macedo, M. L. R. 2019. Biochemical characterization of a Kunitz inhibitor from *Inga edulis* seeds with antifungal activity against *Candida* spp. *Archives of Microbiology*, *201*(2), 223–233. https://doi.org/10.1007/S00203-018-1598-8/METRICS

Efferth, T. 2019. Biotechnology Applications of Plant Callus Cultures. *Engineering*, 5(1), 50–59. https://doi.org/10.1016/J.ENG.2018.11.006

Ershova, N., Sheshukova, E., Kamarova, K., Arifulin, E., Tashlitsky, V., Serebryakova, M., and Komarova, T. 2022. *Nicotiana benthamiana* Kunitz peptidase inhibitor-like protein involved in chloroplast-to-nucleus regulatory pathway in plant-virus interaction. *Frontiers in Plant Science*, 13, 1041867. https://doi.org/10.3389/FPLS.2022.1041867/FULL

Fang, E. F., Bah, C. S. F., Wong, J. H., Pan, W. L., Chan, Y. S., Ye, X. J., and Ng, T. B. 2012. A potential human hepatocellular carcinoma inhibitor from *Bauhinia purpurea* L. seeds: from purification to mechanism exploration. *Archives of Toxicology*, 86(2), 293–304. https://doi.org/10.1007/S00204-011-0751-9

Fang, E. F., Wong, J. H., and Ng, T. B. 2010. Thermostable Kunitz trypsin inhibitor with cytokine inducing, antitumor and HIV-1 reverse transcriptase inhibitory activities from Korean large black soybeans. *Journal of Bioscience and Bioengineering*, 109(3), 211–217. https://doi.org/10.1016/J.JBIOSC.2009.08.483

Farady, C. J., and Craik, C. S. 2010. Mechanisms Of Macromolecular Protease Inhibitors. *Chembiochem : A European Journal of Chemical Biology*, 11(17), 2341. https://doi.org/10.1002/CBIC.201000442

Goldring, J. P. D. 2015. Methods to Concentrate Proteins for Protein Isolation, Proteomic, and Peptidomic Evaluation. *Detection of Blotted Proteins: Methods and Protocols*, 1855:41-59. https://doi.org/10.1007/978-1-4939-2718-0_2

González-Rábade, N., Badillo-Corona, J. A., Aranda-Barradas, J. S., and Oliver-Salvador, M. C. 2011. Production of plant proteases in vivo and in vitro — A review. *Biotechnology Advances*, 29(6), 983–996. https://doi.org/10.1016/J.BIOTECHADV.2011.08.017

Gujjarlapudi, M., Kotarya, B., Mohanraj, S. S., Gupta, D., Prasad, E. R., Kalle, A. M., Jaba, J., Ponnusamy, D., and Padmasree, K. 2023. Development of a rapid process for purification of Bowman-Birk and Kunitz inhibitors from legume seeds, and evaluation of their biophysical, insecticidal, and antimicrobial properties. *International Journal of Biological Macromolecules*, 238, 124050. https://doi.org/10.1016/J.IJBIOMAC.2023.124050

Halan, V., Maity, S., Bhambure, R., and Rathore, A. S. 2019. Multimodal Chromatography for Purification of Biotherapeutics - A Review. *Current Protein and Peptide Science*, 20(1), 4–13. https://doi.org/10.2174/1389203718666171020103559

Huang, T. K., and McDonald, K. A. 2009. Bioreactor engineering for recombinant protein production in plant cell suspension cultures. *Biochemical Engineering Journal*, 45(3), 168–184. https://doi.org/10.1016/J.BEJ.2009.02.008

Karki, U., Fang, H., Guo, W., Unnold-Cofre, C., and Xu, J. 2021. Cellular engineering of plant cells for improved therapeutic protein production. *Plant Cell Reports*, 40(7), 1087–1099. https://doi.org/10.1007/S00299-021-02693-6/TABLES/2

Karray, A., Alonazi, M., Smaoui, S., Michaud, P., Soliman, D., and Bacha, A. Ben. 2020. Purification and Biochemical Characterization of a New Protease Inhibitor from *Conyza dioscoridis* with Antimicrobial, Antifungal and Cytotoxic Effects. *Molecules (Basel, Switzerland)*, 25(22). https://doi.org/10.3390/MOLECULES25225452

Kim, J. Y., Park, S. C., Kim, M. H., Lim, H. T., Park, Y., and Hahm, K. S. 2005. Antimicrobial activity studies on a trypsin–chymotrypsin protease inhibitor obtained from potato. *Biochemical and Biophysical Research Communications*, 330(3), 921–927. https://doi.org/10.1016/J.BBRC.2005.03.057

Kong, X., Li, Y., and Liu, X. 2023. Purification of soybean Kunitz trypsin inhibitor and the mechanism of its passivation by lysine and disulfide bond modifications. *Food Bioscience*, 55, 103042. https://doi.org/10.1016/J.FBIO.2023.103042

Kumar, A., Lodha, S., and Dwiwedi, P. 2015. Production of Serine Protease Inhibitors from *Momordica dioica* Cell Cultures. *Indo Global Journal of Pharmaceutical Sciences*, 05(02), 124–128. https://doi.org/10.35652/IGJPS.2015.13

Kumar, P. P., and Loh, C. S. 2012. Plant tissue culture for biotechnology. *Plant Biotechnology and Agriculture: Prospects for the 21st Century*, 131–138. https://doi.org/10.1016/B978-0-12-381466-1.00009-2

Kumari, V. B. C., Patil, S. M., Ramu, R., Shirahatti, P. S., Kumar, N., Sowmya, B. P., Egbuna, C., Patrick-Iwuanyanwu, K. C., and Uche, C. Z. 2022. Chromatographic techniques: Types, principles, and applications. *Analytical Techniques in Biosciences: From Basics to Applications*, 73–101. https://doi.org/10.1016/B978-0-12-822654-4.00013-0

Kunitz, M. 1945. Crystallization of a Trypsin Inhibitor from Soybean. *Science (New York, N.Y.)*, 101(2635), 668–669. https://doi.org/10.1126/SCIENCE.101.2635.668

López-Otín, C., and Bond, J. S. 2008. Proteases: Multifunctional Enzymes in Life and Disease. *The Journal of Biological Chemistry*, 283(45), 30433. https://doi.org/10.1074/JBC.R800035200

Loyola-Vargas, V. M., and Ochoa-Alejo, N. 2018. An Introduction to Plant Tissue Culture: Advances and Perspectives. *Methods in Molecular Biology (Clifton, N.J.)*, 1815, 3–13. https://doi.org/10.1007/978-1-4939-8594-4_1

Lucena, S. V., Rufino, F. P., Moura, G. E. D. D., Rabêlo, L. M. A., Monteiro, N. K. V., Ferreira, A. T., Perales, J. E. A., Uchôa, A. F., Justo, G. Z., de Oliveira, C. F. R., Migliolo, L., Nader, H. B., Santos, E. A., and Oliveira, A. S. (2022). The Kunitz chymotrypsin inhibitor from *Erythrina velutina* seeds displays activity against HeLa cells through arrest in cell cycle. *3 Biotech*, 12(1). https://doi.org/10.1007/S13205-021-03084-0

Machado, R. J. A., Monteiro, N. K. V., Migliolo, L., Silva, O. N., Pinto, M. F. S., Oliveira, A. S., Franco, O. L., Kiyota, S., Bemquerer, M. P., Uchoa, A. F., Morais, A. H. A., and Santos, E. A. 2013. Characterization and pharmacological properties of a novel multifunctional Kunitz inhibitor from *Erythrina velutina* seeds. *PloS One*, 8(5), e63571. https://doi.org/10.1371/JOURNAL.PONE.0063571

Manzato, V. de M., Di Santo, C., Torquato, R. J. S., Coelho, C., Gallo, G., Hardy, L., Würtele, M., and Tanaka, A. S. 2023. Boophilin D1, a Kunitz type protease inhibitor, as a source of inhibitors for the Zika virus NS2B-NS3 protease. *Biochimie*, 214, 96–101. https://doi.org/10.1016/J.BIOCHI.2023.06.010

Mehmood, S., Imran, M., Ali, A., Munawar, A., Khaliq, B., Anwar, F., Saeed, Q., Buck, F., Hussain, S., Saeed, A., Yasin Ashraf, M., and Akrem, A. 2020. Model prediction of a Kunitz-type trypsin inhibitor protein from seeds of *Acacia nilotica* L. with strong antimicrobial and insecticidal activity. *Turkish Journal of Biology*, 44(4), 188. https://doi.org/10.3906/BIY-2002-20

Mishra, M. 2020. Evolutionary Aspects of the Structural Convergence and Functional Diversification of Kunitz-Domain Inhibitors. *Journal of Molecular Evolution*, 88(7), 537–548. https://doi.org/10.1007/S00239-020-09959-9/FIGURES/4

Novais, C., Molina, A. K., Abreu, R. M. V., Santo-Buelga, C., Ferreira, I. C. F. R., Pereira, C., and Barros, L. 2022. Natural Food Colorants and Preservatives: A Review, a Demand, and a Challenge. *Journal of Agricultural and Food Chemistry*, 70(9), 2789–2805. https://doi.org/10.1021/ACS.JAFC.1C07533

Nowakowski, A. B., Wobig, W. J., and Petering, D. H. 2014. Native SDS-PAGE: High Resolution Electrophoretic Separation of Proteins With Retention of Native Properties Including Bound Metal Ions. *Metallomics : Integrated Biometal Science*, 6(5), 1068. https://doi.org/10.1039/C4MT00033A

Oliva, L. V., Almeida-Reis, R., Theodoro-Junior, O., Oliveira, B. M., Leick, E. A., Prado, C. M., Brito, M. V., Dos Santos Correia, M. T., Paiva, P. M. G., Martins, M. A., Oliva, M. L. V., and Tibério, I. F. L. C. 2015. A plant proteinase inhibitor from *Crataeva tapia* (CrataBL) attenuates elastase-induced pulmonary inflammatory, remodeling, and mechanical alterations in mice. *Process Biochemistry*, 50(11), 1958–1965. https://doi.org/10.1016/J.PROCBIO.2015.06.004

Oliva, M. L. V., Ferreira, R. S., Ferreira, J. G., de Paula, C. A. A., Salas, C. E., and Sampaio, M. U. 2011. Structural and functional properties of kunitz proteinase inhibitors from leguminosae: a mini review. *Current Protein and Peptide Science*, 12(5), 348–357. https://doi.org/10.2174/138920311796391061

Oliva, M. L. V., Silva, M. C. C., Sallai, R. C., Brito, M. V., and Sampaio, M. U. 2010. A novel subclassification for Kunitz proteinase inhibitors from leguminous seeds. *Biochimie*, 92(11), 1667–1673. https://doi.org/10.1016/J.BIOCHI.2010.03.021

Polya, G. M. (2003). Protein and Non-Protein Protease Inhibitors from Plants. *Studies in Natural Products Chemistry*, 29(PART J), 567–641. https://doi.org/10.1016/S1572-5995(03)80015-7

Rajegowda, S. H., and SnehaRani, A. H. 2023. Mechanistic study on the anti-proinflammatory activity of Kunitz type inhibitor from *Caesalpinia decapetala* seeds.

Cell Biochemistry and Function, 41(8), 1133–1146. https://doi.org/10.1002/CBF.3845

Ranasinghe, S. L., Boyle, G. M., Fischer, K., Potriquet, J., Mulvenna, J. P., and McManus, D. P. 2018. Kunitz type protease inhibitor EgKI-1 from the canine tapeworm *Echinococcus granulosus* as a promising therapeutic against breast cancer. *PloS One*, *13*(8), e0200433. https://doi.org/10.1371/JOURNAL.PONE.0200433

Ranasinghe, S., and McManus, D. P. 2013. Structure and function of invertebrate Kunitz serine protease inhibitors. *Developmental and Comparative Immunology*, 39(3), 219–227. https://doi.org/10.1016/J.DCI.2012.10.005

Rani, D., and Vimolmangkang, S. 2024. Production of Recombinant Proteins Using Plant Cell Suspension Cultures and Bioreactor Engineering: A Short Review. *Applications of Plant Molecular Farming.* 141–161. https://doi.org/10.1007/978-981-97-0176-6_6

Rawlings, N. D., Barrett, A. J., Thomas, P. D., Huang, X., Bateman, A., and Finn, R. D. 2018. The MEROPS database of proteolytic enzymes, their substrates and inhibitors in 2017 and a comparison with peptidases in the PANTHER database. *Nucleic Acids Research*, 46(Database issue), D624. https://doi.org/10.1093/NAR/GKX1134

Santos, R. B., Abranches, R., Fischer, R., Sack, M., and Holland, T. 2016. Putting the Spotlight Back on Plant Suspension Cultures. *Frontiers in Plant Science*, 7(MAR2016), 297. https://doi.org/10.3389/FPLS.2016.00297

Simeon, R., and Chen, Z. 2018. In vitro-engineered non-antibody protein therapeutics. *Protein and Cell*, 9(1), 3–14. https://doi.org/10.1007/S13238-017-0386-6

Smith, D., Tikhonova, I. G., Jewhurst, H. L., Drysdale, O. C., Dvořák, J., Robinson, M. W., Cwiklinski, K., and Dalton, J. P. (2016). Unexpected Activity of a Novel Kunitz-type Inhibitor: Inhibition of Cysteine Proteases but not Serine Proteases. *Journal of Biological Chemistry*, 291(37), 19220–19234. https://doi.org/10.1074/JBC.M116.724344

Theodoro-Júnior, O. A., Righetti, R. F., Almeida-Reis, R., Martins-Oliveira, B. T., Oliva, L. V., Prado, C. M., Saraiva-Romanholo, B. M., Leick, E. A., Pinheiro, N. M., Lobo, Y. A., Martins, M. A., Oliva, M. L. V., and Tibério, I. de F. L. C. 2017. A Plant Proteinase Inhibitor from *Enterolobium contortisiliquum* Attenuates Pulmonary Mechanics, Inflammation and Remodeling Induced by Elastase in Mice. *International Journal of Molecular Sciences*, 18(2). https://doi.org/10.3390/IJMS18020403

Thorpe, T. A. 2007. History of plant tissue culture. *Molecular Biotechnology*, 37(2), 169–180. https://doi.org/10.1007/S12033-007-0031-3

Troncoso, F. D., Sánchez, D. S., and Ferreira, M. L. (2022). Production of Plant Proteases and New Biotechnological Applications: An Updated Review. *ChemistryOpen*, 11(3), e202200017. https://doi.org/10.1002/OPEN.202200017

Wang, J., Chitsaz, F., Derbyshire, M. K., Gonzales, N. R., Gwadz, M., Lu, S., Marchler, G. H., Song, J. S., Thanki, N., Yamashita, R. A., Yang, M., Zhang, D., Zheng, C., Lanczycki, C. J., and Marchler-Bauer, A. 2023. The conserved domain database in 2023. *Nucleic Acids Research*, 51(D1), D384–D388. https://doi.org/10.1093/NAR/GKAC1096

Wang, Y., Shi, D., Zou, W., Jiang, Y., Wang, T., Chen, X., Ma, C., Li, W., Chen, T., Burrows, J. F., Wang, L., and Zhou, M. 2024. An Effective Modification Strategy to Build Multifunctional Peptides Based on a Trypsin Inhibitory Peptide of the Kunitz

Family. *Pharmaceutics*, 16(5), 597. https://doi.org/10.3390/PHARMACEUTICS16050597/S1

Williams, A., and Baird, L. G. 2003. DX-88 and HAE: a developmental perspective. *Transfusion and Apheresis Science: Official Journal of the World Apheresis Association: Official Journal of the European Society for Haemapheresis*, 29(3), 255–258. https://doi.org/10.1016/S1473-0502(03)00170-8

Winson, K. W. S., Chew, B. L., Sathasivam, K., and Subramaniam, S. 2020. The establishment of callus and cell suspension cultures of *Hylocereus costaricensis* for the production of betalain pigments with antioxidant potential. *Industrial Crops and Products*, 155, 112750. https://doi.org/10.1016/J.INDCROP.2020.112750

Zhu, M. J., Zhang, G. Q., Wang, H. X., and Ng, T. B. 2011. Isolation and characterization of a Kunitz-type trypsin inhibitor with antiproliferative activity from *Gymnocladus chinensis* (Yunnan bean) seeds. *The Protein Journal*, 30(4), 240–246. https://doi.org/10.1007/S10930-011-9325-1

Zhu-Salzman, K., and Zeng, R. 2015. Insect response to plant defensive protease inhibitors. *Annual Review of Entomology*, 60, 233–252. https://doi.org/10.1146/ANNUREV-ENTO-010814-020816/CITE/REFWORKS

Chapter 6

The Short-Term Liquid Nitrogen Storage of *Pyrostegia venusta* Seeds: A Disruptive *Ex Situ* Conservation Strategy

Mairon César Coimbra[1]
Israel José Pereira Garcia[2]
Hérica de Lima Santos[2]
and Ana Hortência Fonsêca Castro[1,*]

[1]Laboratory of Botany and Medicinal Plants, Postgraduate Program in Biotechnology, Universidade Federal de São João del-Rei (UFSJ), Campus Centro-Oeste Dona Lindu, Divinópolis, MG, Brazil
[2]Laboratory of Cellular Biochemistry, Postgraduate Program in Health Sciences, Universidade Federal de São João del-Rei (UFSJ), Campus Centro-Oeste Dona Lindu, Divinópolis, MG, Brazil

Abstract

Pyrostegia venusta is a plant with recognized medicinal and ornamental uses, and for this reason, it suffers from increasing anthropogenic pressure. The decline in the characteristic habitat for this species encouraged us to search for cryobiotechnological methodologies for *ex situ* conservation since the viability of the seeds is quickly lost. In this study, we evaluated short-term liquid nitrogen (LN) storage of *P. venusta* seeds and how this affected their *in vitro* germination, phenotypic and biochemical characteristics, and production of secondary metabolites in the regenerated seedlings. Short-term storage was carried out using fresh

---

[*] Corresponding Author's Email: acastro@ufsj.edu.br

In: Advances in Biology. Volume 9
Editor: Charles D. Grant
ISBN: 979-8-89530-574-4

seeds with a moisture content of 10.81 ± 0.10% and subjected to different LN exposure times (0, 1, 24, 168, 336, and 504 h). Regarding exposure to LN, seeds cryostored for 7 days (168 h) presented a higher *in vitro* germination percentage (88%) and germination speed index (19.64). However, the phenotypic characteristics of the seedlings obtained were not affected by freezing the seeds. Biochemical analyses revealed that antioxidant metabolism was more pronounced in the aerial parts of seedlings obtained from *in vitro* cultivation. This was not observed for the production and accumulation of secondary metabolites since the levels of phenolic compounds (phenols and flavonoids) were markedly higher in the roots of the analyzed seedlings. Our results reveal that *P. venusta* seeds are tolerant to direct immersion and storage in LN and that these procedures do not significantly affect their physiological quality. We also emphasize that cryopreservation can be an efficient and viable option to preserve seeds and regenerate whole, healthy *P. venusta* plants to obtain bioactive metabolites of interest.

Keywords: cryopreservation, *in vitro* cultivation, medicinal plant, secondary metabolites, oxidative parameters, germination, phenotypic characteristics

Introduction

Pyrostegia venusta (Ker-Gawl.) Miers is a woody vine with dense branches that is native to and found in all Brazilian biomes (Lohmann, 2020). This species can be found dispersed in fields, lining ravines, roadsides, and along pasture fences (Milani et al., 2015). *Pyrostegia vensuta* belongs to the Bignoniaceae family. It is popularly known as "flor-de-São-João" or "cipó-de-São-João", and in popular medicine, it is used to treat coughs, bronchitis, colds, diarrhea, and vitiligo (Rodrigues and Carvalho, 2001; Lorenzi, 2008). There are biological activities including melanogenic, antitumor, vasodilator, anthelmintic, antinociceptive, antioxidant, antimicrobial, and immunomodulatory activities (Veloso et al., 2010, 2012, 2014; Roy et al., 2011; Silva et al., 2011; Moreira et al., 2012, 2015; Nisha et al., 2012; Roy et al., 2012; Reyes-Martínez et al., 2019).

Despite being a plant species with recognized therapeutic and biotechnological potential (Loredo-Carrillo et al., 2013; Braga et al., 2015; Coimbra et al., 2017; Coimbra et al., 2019; Reyes-Martínez et al., 2019) that is dispersed in regions of intense anthropic action (Mostafa et al., 2013; Milani et al., 2015), only two protocols for the *ex situ* conservation of *P. venusta* germplasm have been established (Salomão et al., 2020; Coimbra et al., 2024). In addition to the imminent risk of predatory collection, *P. venusta* has seasonal flowering and fruiting; its seeds have low germination rates and lose viability quickly when stored or in field conditions (Rossato and Kolb, 2010). When brought together and considered from the perspective of promoting sustainable development, these factors justify investment in cryobiotechnological methodologies for the conservation of this species.

Bioprospecting (ethical prospecting) of plant species that produce and accumulate secondary metabolites of economic and/or medicinal interest stands out as a factor of strategic importance for the economies of developing countries or countries that have abundant biodiversity (Dutra et al., 2016). However, the advancement of anthropogenic activities on plant biodiversity can promote the scarcity or extinction of numerous species that have ecological potential and/or pharmaceutical interest even before they are properly studied (Silva et al., 2014). It is in the context of sustainable development that plant biotechnology and its applications in biodiversity conservation have been inserted, especially that of plants with economic and/or therapeutic value. *Ex situ* conservation protocols, such as cryostorage and *in vitro* cultivation of plants that produce bioactive substances of interest, can be innovative and interesting ecological strategies for environmental preservation, in addition to triggering actions with high social and economic impact.

With the unprecedented loss of valuable plant germplasm occurring globally, *ex situ* conservation strategies, such as conserving seeds in germplasm banks, are increasingly important (Liu et al., 2018; Wambugu et al., 2023). A cryobank is an innovative institutional structure for the long-term storage of germplasm derived from *in vitro* cultivation, plant parts (explants), embryos, and seeds under liquid nitrogen (−196°C) or its vapors (−135–180°C). These banks must provide the necessary conditions for the longevity of plant material. To this end, the plant material is stored at low temperatures, which keeps it dormant until used for replanting/cultivation (Nits et al., 2024). Seed storage is certainly the most effective method for the *ex situ* conservation of plant genetic resources, as they are small and therefore occupy a reduced space (Pritchard, 2007; Nits et al., 2024). Seed cryostorage in liquid nitrogen

(LN) was initially developed for the conservation of genetic resources of agriculturally important species (Engelmann, 2004). More recently, cryopreservation has been a viable and safe alternative for the preservation of species at risk of extinction and for biotechnological products (Engelmann, 2011; Wambugu et al., 2023).

Similar to *in vitro* cultivation, cryostorage also imposes a series of stressors on plant material that can induce changes in explants and regenerated plants (Benson, 2008; Engelmann, 2011; Prudente et al., 2019; Coimbra et al., 2024). Therefore, this study focused on reporting the effects of cryopreservation of *P. venusta* seeds on the antioxidant and secondary metabolism of the obtained seedlings as well as on the initial stages of *in vitro* germination. The results described here present a biotechnological option for obtaining plant raw materials that is useful for the pharmaceutical and food industries and offers a viable alternative to the predatory collection of medicinal species.

Method

Plant Material

Pyrostegia venusta (Ker-Gawl.) Miers seeds were collected in Cerrado fragments located in Divinópolis, central-western region of the State of Minas Gerais, Brazil (21°11'36.97" S and 44°55'59.7" W). Fertile samples were collected, herbarized, and identified by researcher Andreia Fonseca Silva from Herário PAMG, belonging to EPAMIG (Agricultural Research Company of Minas Gerais), under registration number PAMG 56307. The collection of botanical material was registered in the Authorization and Information System in Biodiversity (SISBIO) under number 24542-6. This work was registered on the SisGen Platform (Registration A12A940) in accordance with Brazilian Biodiversity Law No. 13,123/2015.

Seed Moisture Level

The moisture content of the seeds was determined using the gravimetric method in an oven at 105 ± 2°C, according to BRASIL (2009).

Short-Term Liquid Nitrogen Storage Trials

Pyrostegia venusta seeds were placed in 5-mL cryotubes and immersed directly in liquid nitrogen (LN) at −196°C for 0, 1, 7, 14, and 21 days (or 0, 1, 24, 168, 336, and 504 h). The cryotubes were removed from LN and thawed directly in water (water bath) for 2 min at 40 ± 1°C (Figueiredo et al., 2017). The cryotubes were taken to the laminar flow hood, where they were opened, and the seeds received aseptic treatment and were inoculated. The experimental design used was completely randomized, consisting of 30 replications per treatment, with each replication consisting of a tube containing a seed.

In Vitro Germination

Pyrostegia venusta seeds were placed in test tubes containing 10 mL of MS medium (Murashige and Skoog, 1962) with a 50% salt concentration, supplemented with 30 g L^{-1} sucrose, solidified with 7 g L^{-1} agar, and pH adjusted to 5.8 ± 0.1 before autoclaving. The material was incubated for 30 days in a growth room at 25 ± 2°C, with a 16-h photoperiod. At the end of the incubation period, the following were evaluated: a) germination speed index (GSI) (Maguire, 1962); b) percentage of germination and contamination; c) number of nodes, buds, and leaflets; d) length of the aerial parts and root; and e) biomass accumulation (fresh and dry mass). The experimental design used was completely randomized, with 30 replications per treatment, with each replication consisting of a tube containing a seed.

Biochemical Analysis

Plantlets obtained from *in vitro* culture for 30 days were selected for biochemical studies and oxidative parameters. The extracts were obtained from 1 g of fresh tissue, which was crushed with the aid of LN in 4 mL of extraction buffer (0.01 M potassium phosphate). This mixture was then centrifuged at 18,000 g for 10 min at 4°C (Mollavali et al., 2016). The supernatant was used to determine the following:

Total Soluble Proteins (PTN). The determination of PTN levels was carried out using the Bradford method (Bradford, 1976). The samples were subjected to analysis using a wavelength of 595 nm. PTN quantification was carried out by constructing a calibration curve using a bovine serum albumin solution as a standard, and the results were expressed in milligrams of PTN per gram of fresh biomass (mg PTN g FB^{-1}).

Total Soluble Sugars (SS). Total SS levels were determined following the methodology described by Yemm and Willis (1954). The samples were subjected to analysis using a wavelength of 620 nm. SS quantification was carried out by constructing a calibration curve using a glucose solution as a standard, and the results were expressed in milligrams of SS per gram of fresh biomass (mg SS g FB^{-1}).

Total Reducing Sugars (RS). The RS levels were determined following the methodology described by Miller (1959). The samples were subjected to analysis using a wavelength of 540 nm. RS quantification was carried out by constructing a calibration curve using a glucose solution as a standard, and the results were expressed in milligrams of RS per gram of fresh biomass (mg RS g FB^{-1}).

Catalase Activity (CAT). To determine the CAT activity, a reaction medium was prepared containing 18 mL of H_2O_2 solution (10% in distilled water), 1 mL of 1M Tris-HCl buffer, pH 8.5 mM EDTA, and 800 μL of distilled water. Then, 995 μL of the reaction medium and 5 μL of the prepared extract were placed in a quartz cuvette (Maehly and Chance, 1954). The reading was carried out using a spectrophotometer at a wavelength of 230 nm for 1 min. The activity calculation was made using the following equation: $(2.3/Dt) \times (a/b) \times (\log A1/A2)$, where a = sample volume in the cuvette; b = total volume of the cuvette; A1 = absorbance value at t=0; and A2 = the absorbance value at the final time, which, in this case, occurred every 15 s during the reaction. All experiments were performed in triplicate, and the results were expressed in units of CAT per milligram of protein (UCAT mg PTN^{-1}).

Reduced Glutathione (GSH). The GSH concentration was determined based on the Ellman reagent reaction, where 5,5'-dithiobis-2-nitrobenzoic acid (DTNB) reacts with the free thiol, generating a more acidic mixed disulfide 2-nitro-5-thiobenzoic acid (Jollow et al., 1974). After 20 min of reaction, the absorbance was read at 412 nm. Values were expressed in milligrams of GSH per gram of fresh biomass (mg GSH g FB^{-1}).

Lipoperoxidation (LP). LP levels were determined by quantifying the thiobarbituric acid reactive substance (TBARS) levels, as described by Kramer et al. (1991). The absorbance was determined at 532 nm, and to obtain the net absorbance values, this value was discounted from the non-specific absorbance at 600 nm. Calculations relating to the release of peroxides through the standard malondialdehyde (MDA) were carried out according to the following equation: {[(A532−A600)/157,000] 106}. The results were expressed in milligrams of MDA per gram of fresh biomass (mg MDA g FB^{-1}).

H_2O_2 Production. The H_2O_2 concentration was determined using the methodology described by Alexieva et al. (2001). H_2O_2 production was calculated from a standard curve using hydrogen peroxide at concentrations from 0 to 100 nM. The samples were subjected to analysis using a wavelength of 390 nm. Values were expressed in milligrams of H_2O_2 per gram of fresh biomass (mg H_2O_2 g FB^{-1}).

Phenylalanine Ammonia-Lyase (PAL) Activity. PAL activity was determined by quantifying the trans-cinnamic acid released from phenylalanine according to the methodology described by Mori et al. (2000). The absorbance was determined at 290 nm, and the absorbance of the control, composed of 50 μM phenylalanine and 100 μL of 100 mM Tris-HCl buffer (pH 8.8), was heated at 37°C for 1.5 h, and 50 μL HCl 6 N was added. The results were expressed in units of PAL per milligram of protein (UPAL mg PTN^{-1}, U = nmol cinnamic acid h^{-1}).

Phytochemical Analysis

Hydroethanolic extracts were prepared from 1 g of dry sample (aerial part and root) and extracted with 10 mL of 70% ethanol by sonication at 35°C for 30 min. The crude extract (CE) was used for phytochemical analysis. All assays were conducted in triplicate.

Total Phenolic Compounds. The content of total phenolic compounds was determined using the Folin–Ciocalteu method described by Slinkard and Singleton (1977), using gallic acid as the standard. The absorbance was measured at 750 nm using a UV-Vis spectrophotometer, and the results were expressed in micrograms of gallic acid equivalents per milligram of crude extract (μg GAEq mg^{-1} CE).

Total Flavonoids. The total flavonoid content was determined using the aluminum chloride method, with rutin as a standard (Brasil, 2010). The absorbance was measured at 425 nm, and the results were expressed in micrograms of rutin equivalents per milligram of crude extract (μg RutEq mg^{-1} CE).

Total Phytosterols. The total phytosterol content was determined using the Liebermann–Buchard method, with β-sitosterol as the reference substance (Araújo et al., 2013). The absorbance was measured at 625 nm, and the results were expressed in micrograms of β-sitosterol equivalents per milligram of crude extract (μg β-sitEq mg^{-1} CE).

Total Alkaloids. The total alkaloid content was determined using the Dragendorff method, as described by Sreevidya and Mehrotra (2003), with allantoin as the standard. The absorbance was measured at 435 nm, and the results were expressed in micrograms of allantoin equivalents per milligram of crude extract (μg AllanEq mg^{-1} CE).

Statistical Analysis

The obtained data were submitted to an analysis of variance (ANOVA) using SISVAR 5.3 (Ferreira, 2014). The results are presented as the mean ± standard deviation. Scott–Knott's multiple range tests were performed to identify significant differences in the data at $p \leq 0.05$.

Results and Discussion

Seed Moisture

The moisture content of *P. venusta* seeds at the time of collection was 10.81 ± 0.10%, which is considered ideal for cryopreservation, making it possible to transfer the seeds directly to LN (Pritchard, 2007; Ferrari et al., 2020). According to Kholina and Voronkova (2012), orthodox wild legume seeds (moisture content between 5 and 12%) are one of the most convenient systems available for long-term storage of genetic information. Thus, the results obtained in this study allow us to infer that *P. venusta* seeds exhibit the characteristics of orthodox seeds since they have a low moisture content (10.81 ± 0.10%), thus enabling them to be cryostored for long periods (Ferrari et al., 2020).

In Vitro Germination and Seedling Growth

The effect of LN exposure on GSI, germination percentage, and subsequent seedling growth parameters was evaluated in *P. venusta*. Table 1 summarizes the data obtained after 30 days of *in vitro* culture. The control group, consisting of seeds not exposed to LN, exhibited a GSI of 9.67, with 56.67% germination and minimal contamination. Seedlings from LN-exposed seeds showed varying responses, depending on the exposure duration. Treatment 3 (7 days of exposure) demonstrated significantly higher germination (73.33%) compared to the control, indicating potential enhancement in seedling establishment following LN exposure. Stress alters the normal course of plant growth and development, metabolism, and other physiological processes (Zevallos et al., 2013). There are many reports indicating that thawing seeds after exposing them to very low temperatures can result in the modification of essential metabolic functions (Prudente et al., 2019). These modifications interfere with the speed and final germination percentage (Kholina and Voronkova, 2012; Gantait et al., 2016).

Morphological parameters, such as number of nodes, buds, and leaflets, were also affected. While no significant differences were observed in these parameters among the treatments, LN exposure generally increased these traits compared to the control, suggesting potential stimulatory effects on early seedling development. Analysis of aerial part (APL) and root length (RL) as well as dry weight (APDW and RDW) indicated subtle variations across treatments (Table 1). Treatment 3 (7 days exposure) showed consistent trends towards increased APL length and APDW compared to the control, although these differences were not statistically significant. Such trends suggest that LN exposure may influence biomass allocation and accumulation, although not significantly altering growth parameters under the conditions tested. The analysis of the phenotypic characteristics of a seedling is an important tool for the preliminary assessment of the effects of cryopreservation (Zevallos et al., 2013). However, few reports have shown phenotypic variations occurring during the *in vitro* cultivation of cryopreserved plant materials. The existence of phenotypic variability is not a rule, as shown by Cejas et al. (2013), who did not observe phenotypic changes in seedlings from cryopreserved seeds of *Phaseolus vulgaris*.

Table 1. Influence of exposure of *Pyrostegia venusta* seeds to LN storage on germination speed index (GSI); germination and contamination percentage; number of nodes, buds, and leaflets; aerial part (APL) and root (RL) length; and dry weight of aerial parts (APDW) and roots (RDW) of seedlings after 30 days of *in vitro* culture

Treat-ment	Exposure to liquid nitrogen	GSI	Germina-tion (%)	Contami nation (%)	Nodes	Buds	Leaflets	APL (cm)	RL (cm)	APDW (mg)	RDW (mg)
Control	-	9.67	56.67^{c}	-	$1.20±0.63^{a}$	$3.00±1.05^{a}$	$10.10±4.38^{a}$	$6.30±1.60^{a}$	$6.05±2.48^{a}$	$15.01±7.24^{a}$	$5.43±3.01^{a}$
1	1 hour	8.34	40.00^{d}	3.33	$0.80±0.63^{a}$	$3.60±1.26^{a}$	$9.60±4.06^{a}$	$5.05±2.06^{a}$	$4.40±1.61^{a}$	$10.61±5.62^{a}$	$3.32±1.61^{a}$
2	1 day (24 hours)	5.76	56.67^{c}	3.33	$1.30±0.82^{a}$	$4.00±1.33^{a}$	$12.20±3.61^{a}$	$6.25±2.23^{a}$	$4.85±2.60^{a}$	$17.92±6.78^{a}$	$3.85±2.53^{a}$
3	7 days (168 hours)	14.39	73.33a	-	$1.20±0.42^{a}$	$4.40±0.84^{a}$	$12.60±1.90^{a}$	$5.75±1.32^{a}$	$5.40±2.39^{a}$	$15.53±5.90^{a}$	$4.50±2.69^{a}$
4	14 days (336 hours)	9.13	53.33^{c}	3.33	$1.10±0.88^{a}$	$3.40±1.35^{a}$	$9.80±3.01^{a}$	$4.95±1.82^{a}$	$6.25±2.25^{a}$	$14.23±7.79^{a}$	$5.03±1.90^{a}$
5	21 days (504 hours)	9.81	63.33^{b}	-	$1.30±1.06^{a}$	$4.00±1.33^{a}$	$11.60±4.20^{a}$	$5.55±2.57^{a}$	$6.15±3.30^{a}$	$17.40±9.24^{a}$	$5.85±3.66^{a}$

[a] Means followed by the same letter in the column are not significantly different ($p > 0.05$), according to Scott- Knott's test.

Table 2. Influence of exposure of seeds to LN on the total soluble protein (PTN), total soluble sugar (SS), and reducing sugar (RS) contents, oxidative metabolism (catalase activity – CAT, reduced glutathione content – GSH, lipoperoxidation levels – LP, hydrogen peroxide production – HPP), and phenylalanine ammonia-lyase (PAL) activity in aerial parts and roots of *Pyrostegia venusta* seedlings after 30 days of *in vitro* culture. The results are presented as the mean ± standard deviation of three repetitions

Treatment	Organ	PTN (mg PTN g FB^{-1})	SS (mg SS g FB^{-1})	RS (mg RS g FB^{-1})	CAT (U CAT mg PTN^{-1})	GSH (mg RG g FB^{-1})	LP (mg MDA g FB^{-1})	HPP (mg H_2O_2 g FB^{-1})	PAL (U PAL mg PTN^{-1})
Control	Aerial Part	257.27±1.51[a]	79.19±1.78[b]	7.12±1.39[c]	1.70±0.08[a]	25.33±1.33[b]	0.49±0.05[b]	0.09[a]	0.50±0.01[a]
	Root	248.07±6.27[a]	86.25±12.89[b]	16.76±0.36[a]	0.52±0.06[c]	31.72±2.74[a]	0.48±0.03[b]	0.06[c]	0.19±0.01[b]
7 days (168 hours)	Aerial Part	262.89±19.29[a]	44.79±1.97[c]	6.55±0.49[c]	1.13±0.12[b]	27.41±3.52[b]	0.57±0.02[a]	0.07[b]	0.46±0.05[a]
	Root	97.90±0.72[b]	126.75±6.76[a]	11.03±0.54[b]	1.08±0.08[b]	12.07±1.10[c]	0.10±0.02[c]	0.03 d	0.40±0.08[a]

[a] Means followed by the same letter in the column are not significantly different ($p > 0.05$), according to Scott- Knott's test.

[b] Note (treatments): Control = Seedlings from embryos not stored in liquid nitrogen; 7 days = Seedlings from embryos stored in liquid nitrogen for 168 hours.

[c] Absence of standard deviation indicates value = 0.00.

Biochemical Responses

Further insights into the biochemical responses are shown in Table 2, which presents data on PTN, SS, and RS contents, oxidative metabolism markers (CAT, GSH, LP, and HPP), and PAL activity in aerial parts and roots after 30 days of culture. LN exposure for 7 days (Treatment 3) led to alterations in protein content and enzymatic activities compared to the control, particularly in roots. This suggests a potential adaptive response to LN exposure, possibly enhancing stress tolerance mechanisms in seedlings. The levels of primary metabolites can be strongly affected by stress conditions and environmental adaptation (Vierstra, 1993; Rosa et al., 2009). Induced thermal stress can cause fractures in biological materials and cause serious damage to stored samples (Panis and Lambardi, 2005). Freezing injury can induce the production of free radicals, mainly reactive oxygen species (Benson, 2008; Reed, 2014). Free radicals attack the lipid fractions of membranes, resulting in the formation of unstable lipid peroxides (Leymarie et al., 2012). Knowledge of biochemical and physiological changes that occur during the *in vitro* growth of plant tissues can provide important information related to the establishment process and, consequently, enable the optimization of conditions for their *in vitro* cultivation and *ex situ* conservation (Prudente et al., 2019; Coimbra et al., 2024).

Secondary Metabolites

The impact of LN exposure on secondary metabolites, such as phenolic compounds, flavonoids, phytosterols, and alkaloids, was evaluated after 30 days of *in vitro* cultivation (Table 3). Treatment 3 (7 days of exposure to LN) resulted in significantly higher phenolic and flavonoid contents in both aerial parts and roots compared to the control, indicating that LN exposure induces metabolic shifts towards increased production of these compounds, which are known for their antioxidant and defense properties. Secondary metabolites are produced in response to the adaptations necessary for the plant's interaction with its environment as well as for its defense (Ramakrishna and Ravishanka, 2011). Considering *in vitro* cultivation, associated with prior seed freezing as a highly stressful condition for *P. venusta* seedlings from these sources, it was expected that the production of secondary metabolites, especially phenolic compounds, would increase. Stress caused by rapid freezing increases the production of phenolic compounds and their subsequent incorporation into the cell wall as suberin or lignin (Griffith and Yaish, 2004). Generally, substances of a phenolic nature are associated with the prevention of oxidative damage due to their ability to capture reactive oxygen species when breaking radical chain reactions during lipid peroxidation (Sakihama and Yamasaki, 2002). Schulz et al. (2016) sought to establish the role of flavonoids in the freezing tolerance of *Arabidopsis thaliana* plants due to the correlation of the abundance of these metabolites and transcripts with freezing tolerance. They found that the majority of flavonols and anthocyanins accumulated after exposure to low temperatures, and transcription factors and enzymes of the flavonoid biosynthetic pathway were encoded. However, evidence for the functional role of flavonoids in plant freezing tolerance has not been demonstrated.

Table 3. Influence of exposure of seeds to LN for 7 days on the total phenolic compound, flavonoid, phytosterol, and alkaloid contents in *Pyrostegia venusta* seedlings cultivated *in vitro* for 30 days. The results are presented as the mean ± standard deviation of three repetitions

Treatment	Organ	Phenolics (µg GAEq mg CE^{-1})	Flavonoids (µg RutEq mg CE^{-1})	Phytosterols (µg β-sisEq mg CE^{-1})	Alkaloids (µg AllanEq mg CE^{-1})
Control	Aerial Part	6.72±0.40[c]	4.84±0.33[b]	16.68±0.73[a]	0.06 d
	Root	5.05±0.46[c]	3.84±0.47[c]	8.03±1.45[b]	0.23±0.01[b]
7 days	Aerial Part	14.44±0.73[b]	4.78±0.29[b]	1.61±0.31[c]	0.15±0.05[c]
(168 hours)	Root	21.09±1.92[a]	7.74±0.43[a]	0.11[c]	0.34±0.05[a]

[a] Means followed by the same letter in the column are not significantly different ($p > 0.05$), according to Scott- Knott's test.

[b] Absence of standard deviation indicates value = 0.00.

[c] Notes (treatments): Control = no exposure to liquid nitrogen; 7 days = 168 hours of exposure to liquid nitrogen.

[d] Legend: GAEq = gallic acid equivalents; RutEq = rutin equivalents; β-sitEq = β-sitosterol equivalents; AllanEq = allantoin equivalents; CE = crude extract.

Conclusion

In this study, the application of short-term LN storage in *P. venusta* seeds was demonstrated as a disruptive conservation tool and a biotechnological alternative for conserving germplasm and obtaining *in vitro* seedlings producing secondary metabolites. In summary, LN exposure of *P. venusta* seeds influenced germination parameters, seedling growth, biochemical responses, and secondary metabolite accumulation. Treatment with LN for 7 days showed promising effects, enhancing germination percentage, seedling biomass, and secondary metabolite content, indicating its potential applications in conservation and propagation strategies for this species. Further investigations into the molecular mechanisms underlying these responses are warranted to fully elucidate the effects of LN exposure on seed physiology and metabolism.

Acknowledgements

We thank to Andreia Fonseca Silva for the botanical identification of this plant. The authors are grateful to the Universidade Federal de São João del-Rei (UFSJ), Conselho Nacional de Desenvolvimento Científico e Tecnológico (CNPq), Coordenacão de Aperfeiçoamento de Pessoal de Nível Superior (CAPES) for financial support. A.H.F.C. are grateful for Productive Fellowship from CNPq (Process 306581/2020-5). This study was financed in part by the Coordenação de Aperfeiçoamento de Pessoal de Nível Superior - Brasil (CAPES), Finance Code 001.

Disclaimer

None

References

Alexieva, V., Sergiev, I., Mapelli, S. and Karanov, E. 2001. The effect of drought and ultraviolet radiation on growth and stress markers in pea and wheat. *Plant Cell and Environment*, 24(12):1337-1344. https://doi.org/10.1046/j.1365-3040.2001.00778.x

Araújo, L.B.D.C., Silva, S.L., Galvão, M.A.M., Ferreira, M.R.A., Araújo, E.L., Randau, K.P. and Soares, L.A.L. 2013. Total phytosterol content in drug materials and extracts from roots of *Acanthospermum hispidum* by UV-VIS spectrophotometry. *Revista Brasileira de Farmacognosia*, 23:736-742. https://doi.org/10.1590/S0102-695X2013000500004

Benson, E.E. 2008. Cryopreservation of phytodiversity: A critical appraisal of theory & practice. *Critical Reviews in Plant Sciences*, 27(3):141-219. https://doi.org/10.1080/07352680802202034

Bradford, M.M. 1976. A rapid and sensitive method for the quantification of microgram quantities of protein utilizing the principle of protein-dye binding. *Analytical Biochemistry*, 72:248-254. https://doi.org/10.1016/0003-2697(76)90527-3

Braga, K.Q., Coimbra, M.C. and Castro, A.H.F. 2015. *In vitro* germination, callus induction and phenolic compounds contents from *Pyrostegia venusta* (Ker Gawl.) Miers. *Acta Scientiarum. Biological Sciences*, 37:151-158. http://dx.doi.org/10.4025/actascibiolsci.v37i2.26067

BRASIL. 2009. *Regras para Análise de Sementes*. Brasília: MAPA/ACS.

BRASIL. 2010. *Farmacopeia Brasileira: Plantas Medicinais*. Brasília: ANVISA.

Cejas, I., Méndez, R., Villalobos, A., Palau, F., Aragón, C., Engelmann, F., Carputo, D., Aversano, R., Martínez, M.E. and Lorenzo, J. C. 2013. Phenotypic and molecular characterization of *Phaseolus vulgaris* plants from non-cryopreserved and cryopreserved seeds. *American Journal of Plant Sciences*, 4:844-849. https://doi.org/10.4236/ajps.2013.44103

Coimbra, M.C., Chagas, R.C.R., Duarte-Almeida, J.M. and Castro, A.H.F. 2017. Influence of plant growth regulators and light on callus induction and bioactive phenolic compounds production in *Pyrostegia venusta* (Bignoniaceae). *Indian Journal of Experimental Biology*, 55(8):584-590.

Coimbra, M.C., Chagas, R.C.R., Vilela, M.S.P. and Castro, A.H.F. 2019. Growth, morphology and bioactive phenolic compounds production in *Pyrostegia venusta* calli. *Biocatalysis and Agricultural Biotechnology*, 18:101036. https://doi.org/10.1016/j.bcab.2019.101036

Coimbra, M.C., Garcia, I.J.P., Santos, H.L. and Castro, A.H.F. 2024. Short-term liquid nitrogen storage of *Pyrostegia venusta* embryos: effects on germination, phenotypic and biochemical characteristics, and in vitro secondary metabolite production. *Horticulturae*, 10(7):695. https://doi.org/10.3390/horticulturae10070695

Dutra, R.C., Campos, M.M., Santos, A.R.S. and Calixto, J.B. 2016. Medicinal plants in Brazil: Pharmacological studies, drug discovery, challenges and perspectives. *Pharmacological Research*, 112:4-29. https://doi.org/10.1016/j.phrs.2016.01.021

Engelmann, F. 2004. Plant cryopreservation: Progress and prospects. *In vitro Cellular & Developmental Biology - Plant*, 40:427-433. https://doi.org/10.1079/IVP2004541

Engelmann, F. 2011. Use of biotechnologies for the conservation of plant biodiversity. *In vitro Cellular & Developmental Biology - Plant*, 47:5-16. https://doi.org/10.1007/s11627-010-9327-2

Ferrari, E.A.P., Colombo, R.C., Faria, R.T. de, Neves, C.S.V.J. and Vero, F. da S. 2020. Degree of moisture in seeds for the cryopreservation of orchids native to Brazil. *Ciência Rural*, 50:e20180498. https://doi.org/10.1590/0103- 8478cr20180498

Ferreira, D.F. 2014. Sisvar: A Guide for its bootstrap procedures in multiple comparisons. *Ciência e Agrotecnologia*, 38(2):109-112. https://doi.org/10.1590/S1413-70542014000200001

Figueiredo, M.A., Coelho, S.V.B., Rosa, S.D.V.F., Vilela, A.L. and Silva, L.C. 2017. Exploratory studies for cryopreservation of *Coffea arabica* L. seeds. *Journal of Seed Science*, 39(2):150-158. https://doi.org/10.1590/2317-1545v39n2171123

Gantait, S., Kundu, S., Wani, S.H. and Das, P.K. 2016. Cryopreservation of forest tree seeds: A mini-review. *Journal of Forest and Environmental Science*, 32:311-322. https://doi.org/10.7747/JFES.2016.32.3.311

Griffith, M. and Yaish, M.W.F., 2004. Antifreeze proteins in overwintering plants: A tale of two activities. *Trends in Plant Science*, 9:399-405. https://doi.org/10.1016/j.tplants.2004.06.007

Jollow, D.J., Mitchell, J.R., Zampaglione, N. and Gillette, J.R. 1974. Bromobenzene-induced liver necrosis. Protective role of glutathione and evidence for 3,4-bromobenzene oxide as the hepatotoxic metabolite. *Pharmacology,* 11(3):151-69. https://doi.org/10.1159/000136485

Kholina, A.B. and Voronkova, N.M. 2012. Seed cryopreservation of some medicinal legumes. *Journal of Botany*, 2012:186891. https://doi.org/10.1155/2012/186891

Kramer, G.F., Norman, H.A., Krizek, D.T. and Mirecki, R.M. 1991. Influence of UV-B radiation on polyamines, lipid peroxidation and membrane lipids in cucumber. *Phytochemistry*, 30(7):2101-2108. https://doi.org/10.1016/0031-9422(91)83595-C

Leymarie, J., Vitkauskaite, G., Hoang, H., Gendreau, E., Chazoule, V., Meimoun, P., Corbineau, F., Hayat, E. and Bailly, C. 2012. Role of reactive oxygen species in the regulation of *Arabidopsis* seed dormancy. *Plant and Cell Physiology*, 53(1):96-106. https://doi.org/10.1093/pcp/pcr129

Liu, U., Breman, E., Cossu, T.A. and Kenney, S. 2018. The conservation value of germplasm stored at the Millennium Seed Bank, Royal Botanic Gardens, Kew, UK. *Biodiversity and Conservation*, 27(6):1347-1386. https://doi.org/10.1007/s10531-018-1497-y

Lohmann, L.G. 2020. "*Pyrostegia* in Flora do Brasil". *Jardim Botânico do Rio de Janeiro*, June 2024. https://floradobrasil2020.jbrj.gov.br/FB113854.

Loredo-Carrillo, S.E., Santos-Diáz, M.L., Leyva, E. and Santos-Diáz, M.S. 2012. Establishment of callus from *Pyrostegia venusta* (Ker Gawl.) Miers and effect of abiotic stress on flavonoids and sterols accumulation. *Journal of Plant Biochemistry and Biotechnology*, 22:312-318. https://doi.org/10.1007/s13562-012-0161-y

Lorenzi, H. 2008. *Plantas Daninhas do Brasil*. 4th ed., Nova Odessa: Plantarum.

Maehly, A.C. and Chance, B. 1954. The assay of catalases and peroxidases. *Methods of Biochemical Analysis*, 1:357-424. https://doi.org/10.1002/9780470110171.ch14

Maguire, J.D. 1962. Speed of germination - Aid in selection and evaluation for seedling emergence and vigor. *Crop Science*, 2(1):176-177. https://doi.org/10.2135/cropsci1962.0011183X000200020033x

Milani, M., Heintze, W., Schafer, G. and De Souza, P.V.D. 2015. Influência do número de folíolos e nós no enraizamento de estacas semilenhosas de cipó-de-São-João. *Ornamental Horticulture*, 21(3):339-344. https://doi.org/10.14295/oh.v21i3.807

Milani, M., Heintze, W., Schafer, G. and De Souza, P.V.D. 2015. Influência do número de folíolos e nós no enraizamento de estacas semilenhosas de cipó-de-São-João. *Ornamental Horticulture*, 21(3):339-344. https://doi.org/10.14295/oh.v21i3.807

Miller, G.L. 1959. Use of dinitrosalicylic acid reagent for determination of reducing sugar. *Analytical Biochemistry*, 31:426-428. https://doi.org/10.1021/ac60147a030

Mollavali, M., Bolandnazar, S.A., Schwarz, D., Rohn, S., Riehle, P. and Nahandi, F.Z. 2016. Flavonol glucoside and antioxidant enzyme biosynthesis affected by mycorrhizal fungi in various cultivars of onion (*Allium cepa* L.). *Journal of Agricultural and Food Chemistry*, 64:71-77. https://doi.org/10.1021/acs.jafc.5b04791

Moreira, C.G., Carrenho, L.Z.B., Pawloski, P.L., Soley, B.S., Cabrini, D.A. and Otuki, M.F. 2015. Pre-clinical evidences of *Pyrostegia venusta* in the treatment of vitiligo. *Journal of Ethnopharmacology*, 168:315-325. https://doi.org/10.1016/j.jep.2015.03.080

Moreira, C.G., Horinouchia, C.D.S., Souza-Filho, C.S., Campos, F.R., Barison, A., Cabrini, D.A. and Otuki, M.F. 2012. Hyperpigmentant activity of leaves and flowers extracts of *Pyrostegia venusta* on murine B16F10 melanoma. *Journal of Ethnopharmacology*, 141:1005-1011. https://doi.org/10.1016/j.jep.2012.03.047

Mori, T., Sakurai, M. and Sakuta, M. 2000. Changes in PAL, CHS, DAHP synthase (DS-Co and Ds-Mn) activity during anthocyanin synthesis in suspension culture of *Fragaria ananassa. Plant Cell, Tissue and Organ Culture*, 62:135-139. https://doi.org/10.1023/A:1026560223795

Mostafa, N.M., El-Dahshan, O. and Singab, A.N.B. 2013. *Pyrostegia venusta* (Ker Gawl.) Miers: A botanical, pharmacological and phytochemical review. *Medicinal & Aromatic Plants*, 2(2):123. https://doi.org/10.4172/2167-0412.1000123

Murashige, T. and Skoog, F. 1962. A revised medium for rapid growth and bioassays with tobacco tissue cultures. *Physiologia Plantarum*, 15(3):473-497. https://doi.org/10.1111/j.1399-3054.1962.tb08052.x

Nisha, P. V., Shruti, N., Swamy, K. S., Kumari, M., Vedamurthy, A. B., Krishna, V. and Hoskeri, J.H. 2012. Anthelmintic activity of *Pyrostegia venusta* using *Pheretima posthuma. International Journal of Pharmaceutical Sciences and Drug Research*, 4(3):205-208.

Nits, O.S., Sementsova, M.V., Osipova, E.S., Tereshonok, D.V. and Gladkov, E.A. 2024. IPPRAS Cryobank for the Conservation of Orthodox Seeds of Rare, Endangered, Medicinal, and Ornamental Plant Species - Current Research. *Plants*, 13(10):1354. https://doi.org/10.3390/plants13101354

Panis, B. and Lambardi, M. "Status of cryopreservation technologies in plants (crops and forest trees)," in *Proceedings of the International Workshop on 'The Role of Biotechnology'*, 2005, pp. 43-54.

Pritchard, H. W. 2007. "Cryopreservation of Desiccation-Tolerant Seeds." In *Cryopreservation and Freeze-Drying Protocols*, edited by Day, J.G., Stacey, G.N., 185-201. New York: Humana Press.

Prudente, D.O., Paiva, R., Domiciano, D., De Souza, L.B., Carpentier S., Swennen, R., Silva, L. C., Nery, F.C., Máximo, W.P.F. and Panis, B. 2019. The cryoprotectant PVS2 plays a crucial role in germinating *Passiflora ligularis* embryos after cryopreservation by influencing the mobilization of lipids and the antioxidant metabolism. *Journal of Plant Physiology*, 239:71-82. https://doi.org/10.1016/j.jplph.2019.05.014

Ramakrishna, A. and Ravishankar, G.A., 2011. Influence of abiotic stress signals on secondary metabolites in plants. *Plant Signaling & Behavior,* 6(11):1720-1731. https://doi.org/10.4161/psb.6.11.17613

Reed, B.M. 2014. Antioxidants and cryopreservation, the new normal? *Acta Horticulturae*, 1039:41-48. https://doi.org/10.17660/ActaHortic.2014.1039.3

Reyes-Martínez, A., Valle-Aguilera, J.R., Antunes-Ricardo, M., Gutiérrez-Uribe, J., Gonzalez, C. and Santos-Díaz, M.S. 2019. Callus from *Pyrostegia venusta* (Ker Gawl.) Miers: A source of phenylethanoid glycosides with vasorelaxant activities. *Plant Cell, Tissue and Organ Culture,* 139(1): 119-129, 2019. https://doi.org/10.1007/s11240-019-01669-5

Rodrigues, V.E.G. and Carvalho, D.A. 2007. Levantamento etnobotânico de plantas medicinais no domínio dos cerrados na região do Alto Rio Grande – Minas Gerais. *Revista Brasileira de Plantas Medicinais*, 9(2):17-35.

Rosa, M., Prado, C., Podazza, G., Interdonato, R., González, J.A., Hilal, M. and Prado, F.E. 2009. Soluble sugars-Metabolism, sensing and abiotic stress. *Plant Signaling & Behavior*, 4(5):388-393. https://doi.org/10.4161%2Fpsb.4.5.8294

Rossato, D.R. and Kolb, R.M. 2010. Germinação de *Pyrostegia venusta* (Bignoniaceae), viabilidade de sementes e desenvolvimento pós-seminal. *Brazilian Journal of Botany*, 33(1):51-60. https://doi.org/10.1590/S0100-84042010000100006

Roy, P., Amdekar, S., Kumar, A. and Singh, V. 2011. Preliminary study of the antioxidant properties of flowers and roots of *Pyrostegia venusta* (Ker Gawl.) Miers. *BMC Complementary and Alternative Medicine*, 11:69-77. https://doi.org/10.1186/1472-6882-11-69

Roy, P., Amdekar, S., Kumar, A., Singh, R., Sharma, P. and Singh, V. 2012. *In vivo* antioxidative property, antimicrobial and wound healing activity of flower extracts of *Pyrostegia venusta* (Ker Gawl.) Miers. *Journal of Ethnopharmacology*, 140:186-192. https://doi.org/10.1016/j.jep.2012.01.008

Sakihama, Y. and Yamasaki, H., 2002. Lipid Peroxidation induced by phenolics in conjunction with aluminum ions. *Biologia Plantarum*, 45(2):249-254. https://doi.org/10.1023/A:1015152908241

Salomão, A.N., Santos, I.R.I. and José, S.C.B.R. 2020. Cryopreservation of *Pyrostegia venusta* (Ker Gawl.) Miers seeds. *Hoehnea*, 47:e1042019. http://dx.doi.org/10.1590/2236-8906-104/2019

Schulz, E., Tohge, T., Zuther, E., Fernie, A.R. and Hincha, D.K., 2016. Flavonoids are determinants of freezing tolerance and cold acclimation in *Arabidopsis thaliana*. *Scientific Reports*, 6:34027. https://doi.org/10.1038/srep34027

Silva, L.C., Paiva R., Swennen, R., Andrè, E., and Panis, B. 2014. Cryopreservation of *Byrsonima intermedia* embryos followed by room temperature thawing. *Acta Scientiarum. Agronomy*, 36(3):309-315. https://doi.org/10.4025/actasciagron.v36i3.19688

Silva, P.B., Medeiros, A.C.M., Duarte, M.C.T., Ruiz, A.L.T.G., Kolb, R.M., Frei, F. and Santos, C. 2011. Avaliação do potencial alelopático, atividade antimicrobiana e antioxidante dos extratos orgânicos das folhas de *Pyrostegia venusta* (Ker Gawl.) Miers (Bignoniaceae). *Revista Brasileira Plantas Medicinais*, 13(4):447-455. https://doi.org/10.1590/S1516-05722011000400010

Slinkard, K. and Singleton, V.L. 1977. Total phenol analyses: Automation and comparison with manual methods. *American Journal of Enology and Viticulture*, 28(1):49-55. https://doi.org/10.5344/ajev.1977.28.1.49

Sreevidya, N. and Mehrotra, S. 2003. Spectrophotometric method for estimation of alkaloids precipitable with Dragendorff's reagent in plant materials. *Journal of AOAC International*, 86(6):1124-1127. https://doi.org/10.1093/jaoac/86.6.1124

Veloso, C.C., Bitencourt, A.D., Cabral, L.D. M., Franqui, L.S., Dias, D.F., Santos, M.H., Soncini, R. and Giusti-Paiva, A. 2010. *Pyrostegia venusta* attenuate the sickness behavior induced by lipopolysaccharide in mice. *Journal of Ethnopharmacology*, 132:355-358. https://doi.org/10.1016/j.jep.2010.07.053

Veloso, C.C., Cabral, L.D.M., Bitencourt, A.D., Franqui, L.S., Santa-Cecília, F.V., Dias, D.F., Soncini, R., Vilela, F.C. and Giusti-Paiva, A. 2012. Anti-inflammatory and antinociceptive effects of the hydroethanolic extract of the flowers of *Pyrostegia venusta* in mice. *Revista Brasileira de Farmacognosia*, 22(1):162-168. https://doi.org/10.1590/S0102-695X2011005000209

Veloso, C.C., Oliveira, M.C., Oliveira, C.C., Rodrigues, V.G., Giusti-Paiva, A., Teixeira, M.M., Duarte, I.D., Ferreira, A.V. and Perez, A.C. 2014. Hydroethanolic extract of *Pyrostegia venusta* (Ker Gawl.) Miers flowers improves inflammatory and metabolic dysfunction induced by high-refined carbohydrate diet. *Journal of Ethnopharmacology*, 151:722-728. https://doi.org/10.1016/j.jep.2013.11.046

Vierstra, R.D. 1993. Protein degradation in plants. *Annual Review of Plant Physiology and Plant Molecular Biology*, 44:385-410. https://doi.org/10.1146/annurev.pp.44.060193.002125

Wambugu, P.W., Nyamongo, D.O. and Kirwa, E.C. 2023. Role of Seed Banks in Supporting Ecosystem and Biodiversity Conservation and Restoration. *Diversity*, 15(8): 896. https://doi.org/10.3390/d15080896

Yemm, E.W. and Willis, A.J. 1954. The estimation of carbohydrates in plant extracts by anthrone. *The Biochemical Journal*, 57:508-514. https://doi.org/10.1042%2Fbj0570508

Zevallos, B., Cejas, I., Rodríguez, R.C., Yabor, L., Aragón, C., González, J., Engelmann, F., Martínez, M.E. and Lorenzo, J.C. 2013. Biochemical characterization of ecuadorian wild *Solanum lycopersicum* MILL. plants produced from non-cryopreserved and cryopreserved seeds. *CryoLetters*, 34(4):413-421.

Chapter 7

The Overlap of Hypoparathyroidism and Ankylosing Spondylitis: A Clinical and Radiological Investigations

Taieb Ach[1,3,*]
Nejla Amri[2]
and Asma Ben Abdelkrim[1]

[1]Department of Endocrinology and Diabetology, University Hospital of Farhat Hached, Sousse, Tunisia
[2]Department of Rheumatology, University Hospital of Farhat Hached, Sousse, Tunisia
[3]Laboratory of Exercise Physiology and Pathophysiology; L.R.19ES09, Tunisia

Abstract

Idiopathic hypoparathyroidism is a rare condition, either inherited or acquired, characterized by insufficient secretion of parathyroid hormone (PTH). This deficiency leads to varied clinical manifestations, primarily due to neuromuscular irritability, and often presents with skeletal abnormalities. These abnormalities can closely mimic those observed in patients with ankylosing spondylitis (AS), both clinically and radiologically.

In this report, we present a case of a patient with idiopathic hypoparathyroidism who was misdiagnosed with AS. For years, the patient received anti-inflammatory treatment for chronic inflammatory back pain, under the incorrect diagnosis of AS. This case highlights the importance of thoroughly investigating potential mineral metabolism disturbances in patients presenting with symptoms of inflammatory diseases.

---

[*] Corresponding Author's Email: achtaieb@gmail.com

In: Advances in Biology. Volume 9
Editor: Charles D. Grant
ISBN: 979-8-89530-574-4

The delayed diagnosis of hypoparathyroidism in this patient underscores the clinical challenge in distinguishing hypoparathyroid-related spondyloarthropathy from AS. Accurate differentiation between these two conditions is crucial, as their management strategies differ significantly. While AS treatment primarily involves anti-inflammatory medications and physical therapy, hypoparathyroidism requires a targeted approach to correct the underlying hormonal deficiency.

This case emphasizes the necessity for healthcare professionals to maintain a high index of suspicion for endocrine disorders in patients with atypical presentations of inflammatory back pain. Comprehensive metabolic evaluation should be considered to avoid misdiagnosis and ensure appropriate treatment. By recognizing and addressing the specific needs of patients with hypoparathyroidism, we can prevent unnecessary treatments and improve clinical outcomes.

Keywords: hypoparathyroidism, hypocalcemia, parathyroid hormone, ankylosing spondylitis

Abbreviations

IHP: Idiopathic hypoparathyroidism
HP: hypoparathyroidism
PTH: parathyroid hormone
AS: ankylosing spondylitis

Introduction

Hypoparathyroidism (HP) is an endocrine deficiency disorder characterized by insufficient production or action of parathyroid hormone (PTH), leading to disturbances in calcium and phosphate homeostasis. This condition can be either acquired or inherited. Acquired forms are most commonly due to surgical complications, particularly after thyroid or parathyroid surgery, or secondary to severe hypomagnesemia [1]. Congenital forms can result from genetic mutations affecting parathyroid gland development or function. When no clear etiology is identified, the condition is termed idiopathic hypoparathyroidism (IHP) [2].

HP is a rare disorder with an estimated prevalence of 24 to 37 per 100,000 individuals [3]. The acquired form is the most prevalent, particularly in adults, due to the higher incidence of thyroid surgeries. Congenital and idiopathic forms are less common and can present at any age [4].

HP has a broad spectrum of clinical manifestations, varying from acute to chronic symptoms. Acute symptoms include neuromuscular irritability, such as tetany, muscle cramps, carpopedal spasm, and seizures, primarily due to hypocalcemia [5]. Cardiac symptoms may involve prolonged QT intervals on electrocardiograms, arrhythmias, and in severe cases, heart failure. Respiratory complications like laryngospasm and bronchospasm can occur, leading to acute respiratory distress. Chronic symptoms include renal complications such as nephrocalcinosis, nephrolithiasis, and impaired renal function due to chronic hyperphosphatemia [6]. Neurological symptoms may involve cognitive impairment, basal ganglia calcifications, and extrapyramidal symptoms, while neuropsychiatric symptoms can include anxiety, depression, and mood disorders. Skeletal manifestations can present as osteosclerosis, dental abnormalities, and paravertebral calcification, leading to pain, stiffness, limited movement, and a posture resembling ankylosing spondylitis (AS) [7].

Diagnosis of HP is based on clinical presentation, biochemical findings, and imaging studies. Biochemical findings typically show low serum calcium, low PTH levels, high serum phosphate, and the need to check magnesium levels to rule out hypomagnesemia [8]. Imaging studies such as neck ultrasound, skeletal X-rays, and CT or MRI scans are used to assess the parathyroid glands, detect osteosclerosis and soft tissue calcifications, and identify intracranial abnormalities. Genetic testing is indicated in cases of congenital or idiopathic hypoparathyroidism to identify potential genetic mutations [9].

The primary defect in HP is the insufficient secretion or action of PTH, leading to disrupted calcium and phosphate homeostasis. PTH normally increases calcium reabsorption in the kidneys, promotes calcium release from bones, and enhances intestinal calcium absorption by activating vitamin D. In HP, these mechanisms are impaired, resulting in hypocalcemia and hyperphosphatemia [10].

Management of HP involves addressing both acute and chronic aspects. Acute management may include intravenous calcium to rapidly correct severe hypocalcemia. Chronic management typically involves oral calcium and active vitamin D analogs to maintain normocalcemia, along with regular monitoring of serum calcium, phosphate, and magnesium levels to prevent complications [11].

On the other hand, Spondyloarthropathy (SpA) encompasses a diverse group of rheumatic diseases characterized by common clinical and genetic features. These diseases are categorized as either peripheral or axial (axSpA), depending on the primary areas affected. Ankylosing spondylitis (AS), a subtype of SpA, is an autoimmune condition primarily affecting the spine, sacroiliac joints (SIJs), and adjacent soft tissues such as tendons and ligaments. In advanced stages, inflammation can lead to fibrosis and calcification, resulting in reduced flexibility and spinal fusion, often described as a "bamboo spine." Key clinical manifestations include back pain, progressive spinal rigidity, and inflammation of the hips, shoulders, peripheral joints, and fingers or toes. Extra-articular manifestations, such as acute anterior uveitis and inflammatory bowel disease (IBD), also occur, though their prevalence differs between East Asian and Caucasian populations. For example, a study of 988 AS patients in East Asia found a 0.4% prevalence of IBD, while Western studies report approximately 5% to 10% of AS patients presenting with IBD.

The prevalence of AS is closely related to the presence of the human leukocyte antigen (HLA)-B27 in specific populations. Research shows that AS prevalence is about 5% to 6% in HLA-B27-positive groups. A 2009 national survey in the United States revealed varying prevalence rates of HLA-B27 positivity across ethnic communities, with 7.5% in non-Hispanic whites, 4.6% in Mexican-Americans, and 1.1% in non-Hispanic blacks. Males are significantly more affected by AS compared to females, although the incidence is similar between men and women in nonradiographic axial spondyloarthropathy (nr-axSpA), a condition meeting clinical criteria for axSpA without radiological evidence of sacroiliitis. A meta-analysis of eight studies involving 2,236 AS patients and 1,242 nr-axSpA patients found that males represented 70.4% of AS cases and 46.5% of nr-axSpA cases. Genetic susceptibility studies have identified recurrent risk factors across different generations of relatives: monozygotic twins (63%), first-generation relatives (8.2%), second-generation relatives (1.0%), and third-generation relatives (0.7%).

To assist in diagnosis, several criteria sets are used in clinical practice. The modified New York (mNY) criteria are the most widely applied for diagnosing AS, requiring at least one clinical criterion and radiological evidence. Other classification systems include the Amor criteria and the European Spondyloarthropathy Study Group (ESSG) criteria. In 2012, AV Tubergen reviewed these classification criteria, and the ASAS criteria, a prominent set for diagnosing axSpA, have gained traction in Europe with the

2016 update of the ASAS-EULAR management recommendations for axial spondyloarthritis. To aid in differentiating AS from other forms of SpA, Joel D. Taurog developed an algorithm for the diagnosis or exclusion of axSpA.

In this report, we present a case of IHP that was initially misdiagnosed as AS, leading to inappropriate anti-inflammatory treatment. This case underscores the critical importance of evaluating underlying disturbances in mineral metabolism in patients presenting with inflammatory conditions.

Observation

A 61-year-old man was admitted to the hospital for generalized seizures accompanied by clonic spasms and contractures. His medical history included 20 years of chronic inflammatory cervical and lower back pain, with morning stiffness that was resistant to non-steroidal anti-inflammatory drugs. He had undergone bilateral cataract surgery five years earlier. The patient reported no family history of endocrinologic or rheumatologic diseases, chronic alcoholism, or history of cervical surgery or irradiation.

On examination, the patient was found to be irritable, with desquamating dry skin, brittle nails, thin and dry hair, and streaked tooth enamel. He exhibited extrapyramidal syndrome, characterized by akinesia, resting tremor, plastic rigidity, and cogwheel phenomenon. Schovstek's and Trousseau's signs were negative.

His posture was typical for ankylosing spondylitis, with cervical anteflexion, dorsal kyphosis, loss of lumbar lordosis, straightening of the thoracic and lumbar spine, and compensatory hip flexion. The occiput-to-wall distance was 25 cm, the chin-to-sternum distance was 10 cm, and Schober's test could not be performed. There was no evidence of peripheral arthritis.

Laboratory investigations revealed normal plasma glucose levels, hypocalcemia (serum calcium: 1.1 mmol/L [normal range: 2-2.5 mmol/L]), and corrected calcium of 1.59 mmol/L (albumin: 25.08 g/L [normal range: 40.2-47.6 g/L]). Hyperphosphatemia was present (2.2 mmol/L [normal range: 0.81-1.36 mmol/L]), along with hypomagnesemia (0.7 mmol/L [normal range: 0.74-0.9 mmol/L]). PTH levels were low at 5.5 pg/mL (normal range: 10-65 pg/mL), and vitamin D deficiency was noted (15 ng/mL). Renal function was normal (creatinine: 81 μmol/L [normal range: 70-130 μmol/L]), and hypocalciuria was present (urinary calcium: 1.49 mmol/24 hours). The sedimentation rate was elevated at 104 mm/hour, and there was no anemia.

The Synacthen test showed a cortisol peak at 290 ng/mL, and antithyroid antibodies were negative.

An electrocardiogram revealed a prolonged QT interval. Brain computed tomography showed findings consistent with Fahr syndrome. A spine radiograph demonstrated multiple osseous excrescences forming inter-somatic bridges in the cervical and lumbar spine (Figure 1). A pelvic X-ray revealed no evidence of sacroiliitis but showed bilateral ossification at the superior acetabular margin (Figure 2).

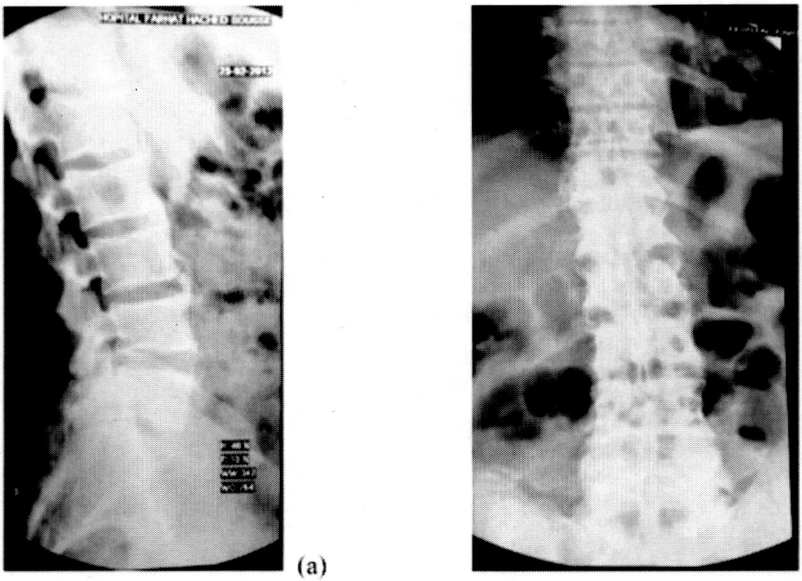

(a) (b)

Figure 1. Lateral (a) and face (b) lumbar spine radiograph showing inter-somatic bridges.

The diagnosis of idiopathic hypoparathyroidism (IHP) with an ankylosing spondylitis (AS)-like presentation was confirmed. The patient was initiated on treatment with calcium supplements and vitamin D derivatives, resulting in significant clinical and laboratory improvements.

Upon starting treatment, the patient's serum calcium levels gradually normalized, rising from 1.1 mmol/L to within the normal range. Concurrently, his hyperphosphatemia improved, and serum phosphate levels decreased to within the normal limits. His PTH levels remained low but stabilized, reflecting a controlled management of his condition. Vitamin D levels were corrected, which facilitated the proper absorption of calcium and contributed to the overall biochemical balance.

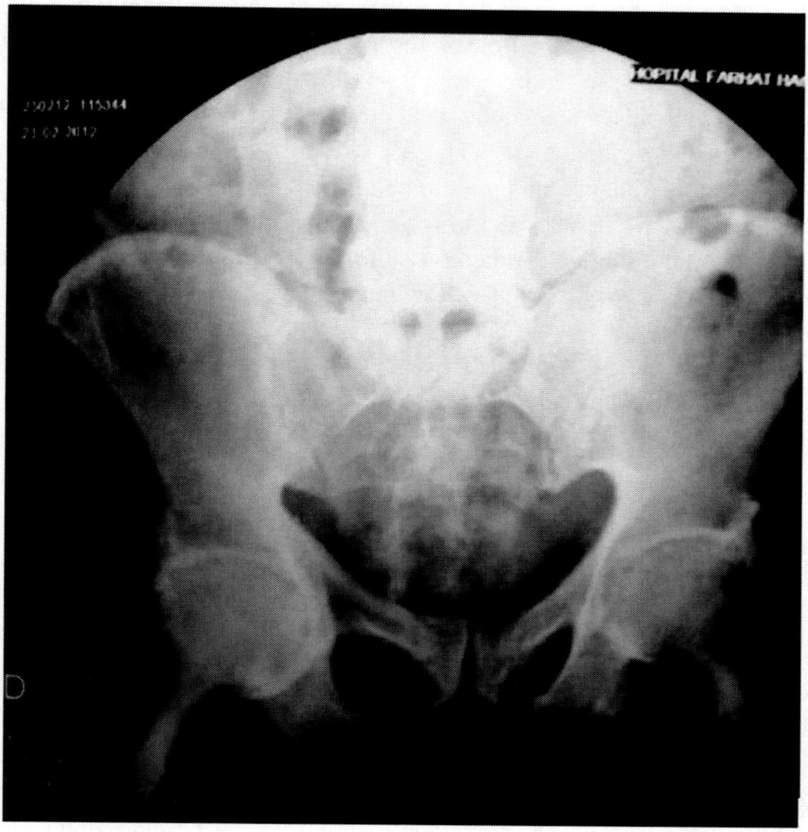

Figure 2. Pelvic radiograph showing sacro-iliac joint integrity and bilateral ossification at the superior acetabular margin.

Clinically, the patient experienced a marked reduction in neuromuscular irritability. The generalized seizures and clonic spasms ceased, and his irritability diminished. There was also a noticeable improvement in his skin condition, with reduced dryness and desquamation, and his nails and hair showed signs of becoming less brittle and healthier over time.

The patient's posture, which had been reminiscent of ankylosing spondylitis, showed gradual improvement. He regained some of his spinal mobility, with reduced pain and stiffness in the cervical and lumbar regions. Although the occiput-to-wall distance and chin-to-sternum distance remained abnormal, there was a measurable decrease, indicating improvement. Schober's test, previously impossible to perform, became feasible, albeit with limited flexibility.

Radiological follow-up indicated that while the pre-existing osseous excrescences and inter-somatic bridges in the cervical and lumbar spine remained, there was no further progression of these skeletal abnormalities. This suggested a halt in the pathological process associated with his previous state of untreated hypoparathyroidism.

Additionally, the patient's renal function remained stable, and there was no development of nephrocalcinosis or nephrolithiasis, as confirmed by follow-up imaging studies. His neuropsychiatric symptoms, including anxiety and depression, showed improvement with the stabilization of his calcium and phosphate levels.

Overall, the management of IHP with calcium and vitamin D derivatives not only corrected the biochemical abnormalities but also significantly enhanced the patient's quality of life and physical functioning.

Discussion

The presence of hypocalcemia, hyperphosphatemia, and low levels of parathyroid hormone confirmed the diagnosis of hypoparathyroidism in this patient [15]. Hypoparathyroidism is an endocrine disorder characterized by inadequate secretion or action of parathyroid hormone (PTH), resulting in hypocalcemia and hyperphosphatemia. This condition can be caused by various congenital or acquired disorders [16].

The most common cause of hypoparathyroidism is iatrogenic, typically occurring after anterior neck surgery, such as thyroidectomy or parathyroidectomy. Other acquired causes include autoimmune diseases, genetic abnormalities, and destructive or infiltrative disorders of the parathyroids, such as hemochromatosis, Wilson's disease, or metastatic cancer [17]. Hypoparathyroidism can also result from impaired PTH secretion due to hypomagnesemia, although this was not the case in our patient [18].

Inherited forms of hypoparathyroidism include conditions like DiGeorge syndrome, which involves congenital absence or hypoplasia of the parathyroid glands due to 22q11.2 deletion, and familial isolated hypoparathyroidism, caused by mutations in the PTH gene or other related genes [19]. Idiopathic hypoparathyroidism (IHP) refers to cases where the etiology remains unknown despite thorough investigation [20].

In our patient, the idiopathic origin of hypoparathyroidism was established due to the absence of a history of parathyroid surgery, radiation exposure, or familial autoimmune polyendocrinopathies. Additionally, normal magnesium levels and the lack of genetic abnormalities supported the diagnosis of IHP [21].

The presence of bilateral cataracts and Fahr syndrome, characterized by intracranial calcifications, indicated a long-standing history of hypoparathyroidism. This condition is known to cause soft-tissue ectopic calcifications, diffuse ligamentous, and entheseal ossifications [22]. Prolonged hypoparathyroidism can lead to ossification of spinal ligaments and the formation of bridging osteophytes. These changes can mimic other conditions such as spondyloarthropathies and diffuse idiopathic skeletal hyperostosis (DISH) [23].

Our patient exhibited clinical features suggestive of ankylosing spondylitis (AS), including cervical anteflexion, dorsal kyphosis, and loss of lumbar lordosis. However, radiologic findings did not support this diagnosis. The intervertebral bone bridges observed in our patient were exuberant, whereas syndesmophytes in AS are typically thin and follow a vertical path [24]. Additionally, the sacroiliac joints were preserved, which is not characteristic of AS, where sacroiliitis is a hallmark and a major diagnostic criterion [25].

In contrast, hypoparathyroidism generally spares the sacroiliac joints, differentiating it from AS. This distinction is crucial for accurate diagnosis and appropriate treatment. In hypoparathyroidism, the skeletal manifestations, including osteosclerosis and ossification of spinal ligaments, can be mistaken for other rheumatologic conditions [26].

Furthermore, the chronic complications of hypoparathyroidism extend beyond the skeletal system. Renal complications, such as nephrocalcinosis and nephrolithiasis, can arise due to chronic hyperphosphatemia and impaired calcium-phosphate metabolism [27]. Neurological and neuropsychiatric symptoms, including cognitive impairment, basal ganglia calcifications, and mood disorders, are also common [28]. These systemic effects highlight the importance of comprehensive management and monitoring of patients with hypoparathyroidism [29].

The treatment of hypoparathyroidism focuses on correcting hypocalcemia and maintaining normal calcium and phosphate levels. This typically involves the administration of oral calcium supplements and active vitamin D analogs, such as calcitriol, to enhance intestinal calcium absorption. In some cases, recombinant PTH (teriparatide) may be used, especially in patients who do not respond adequately to conventional therapy [30]. Regular monitoring of serum calcium, phosphate, magnesium, and renal function is essential to prevent complications and ensure effective management [31].

Ankylosing spondylitis (AS) is a chronic, inflammatory disease primarily affecting the axial spine, manifesting with a range of clinical signs and symptoms. The hallmark features of the condition include chronic back pain and progressive spinal stiffness. AS involves the spine, sacroiliac (SI) joints, peripheral joints, digits, and entheses. It often leads to impaired spinal mobility and can result in postural abnormalities. Patients can also experience buttock and hip pain. Peripheral arthritis, enthesitis, and dactylitis ("sausage digits") are all associated with AS. In addition to skeletal involvement, it can affect various organs outside the joints, including inflammatory bowel disease (affecting up to 50% of individuals), acute anterior uveitis (seen in 25%-35% of cases), and psoriasis (approximately 10% occurrence) [31].

AS is linked to an increased risk of cardiovascular disease, believed to stem from the systemic inflammation present in individuals with AS. Pulmonary complications are also associated with AS, as diminished chest wall expansion and limited spinal mobility can predispose individuals to a restrictive pulmonary pattern [32].

In suspected cases of AS, a comprehensive evaluation of the entire body is recommended due to the systemic nature of the disease and its potential involvement in multiple organ systems. Back pain is a common complaint among patients, and the characteristic type of back pain in AS is "inflammatory" in nature. Inflammatory back pain is identified by at least four of the following five features: onset of symptoms before the age of 40, gradual and insidious onset, relief with exercise, lack of improvement with rest, and nocturnal pain with improvement upon arising [33]. Additionally, patients often exhibit spinal stiffness, limited mobility, and postural changes, particularly hyperkyphosis.

During the evaluation, the patient's history and physical examination should encompass all body systems, as AS can present with both axial and peripheral musculoskeletal symptoms as well as extra-articular features. A detailed medical history should be obtained to identify or rule out associated conditions such as psoriasis, inflammatory bowel disease, and uveitis, which may be correlated with AS [34].

Laboratory findings in ankylosing spondylitis (AS) are typically nonspecific but can provide supportive evidence for diagnosis. Approximately 50% to 70% of patients with active AS exhibit elevated levels of acute phase reactants, such as erythrocyte sedimentation rate (ESR) and C-reactive protein

(CRP). However, normal ESR and CRP levels should not be used to exclude the possibility of AS [35]. Several imaging abnormalities, particularly those affecting the spine and sacroiliac (SI) joints, are associated with AS. Evidence of sacroiliitis on imaging, whether radiographic or magnetic resonance imaging (MRI), is considered a major inclusion criterion for AS according to the Assessment of Spondyloarthritis International Society (ASAS) 2009 axial spondyloarthritis criteria [36].

A standardized plain radiographic grading scale exists for sacroiliitis, ranging from normal (0) to most severe (IV), as follows:

- 0: Normal SI joint width, sharp joint margins
- I: Suspicious
- II: Sclerosis, some erosions
- III: Severe erosions, pseudo dilation of the joint space, partial ankylosis
- IV: Complete ankylosis

In the initial years of AS, plain radiographic changes in the SI joints can be subtle; however, these changes usually become more evident over the first decade of the disease. The most noticeable abnormalities on radiographs include subchondral erosions, sclerosis, and joint fusion, which are typically symmetrical, affecting both sides of the SI joints similarly. Throughout AS, distinct radiographic changes can progressively develop. In the early stages, a notable sign is the "squaring" of vertebral bodies, best visualized on lateral X-rays. This squaring occurs due to inflammation and bone deposition, resulting in the loss of the normal concavity of the anterior and posterior borders of the vertebral body. Additionally, early-stage radiographs may reveal Romanus lesions, or "shiny corner signs," characterized by small erosions and reactive sclerosis at the corners of the vertebral bodies [37].

Late-stage findings on radiographs include ankylosis (fusion) of the facet joints of the spine, the presence of syndesmophytes, and calcification of the anterior longitudinal ligament, supraspinous ligaments, and interspinous ligaments. This calcification may be seen on imaging as the "dagger sign," appearing as a single radiodense line vertically running down the spine on frontal radiographs. The classic radiographic finding in late-stage AS is the "bamboo spine sign," which refers to vertebral body fusion by syndesmophytes. The bamboo spine typically involves the thoracolumbar or lumbosacral junctions, leading to progressive back stiffness [38]. While plain radiography is the first-line imaging modality in AS, additional imaging with MRI may be necessary to detect more subtle abnormalities, such as fatty or inflammatory changes [39]. MRI can reveal active inflammatory lesions in the SI joints, appearing as bone marrow edema (BME) on short tau inversion recovery (STIR) and T2-weighted images with fat suppression. It should be noted that BME on MRI can also be seen in up to 23% of patients with mechanical back pain and 7% of healthy individuals [40].

In our case, the idiopathic origin was retained given the absence of a previous history of parathyroid surgery, radiation exposure, or familial autoimmune polyendocrinopathies, the normality of the magnesium level, and the absence of abnormalities leading to a genetic cause [41]. The presence of cataracts and Fahr syndrome testifies to the long evolution of hypoparathyroidism. This disease is characterized by soft-tissue ectopic calcifications, diffuse ligamentous and entheseal ossifications. In long-lasting hypoparathyroidism, there is also ossification of spinal ligaments and bridging osteophytes [42]. Clinical and radiologic changes suggestive of spondyloarthropathy and diffuse idiopathic skeletal hyperostosis might be seen [30, 43]. Our patient had a clinical presentation of ankylosing spondylitis. However, radiologic findings did not comply with this diagnosis. In fact, in his case, the intervertebral bone bridges were exuberant while syndesmophytes in AS are thin and have a vertical path [44]. Moreover, the sacroiliac joint was preserved while sacroiliitis is the hallmark of AS and constitutes a major criterion for this diagnosis [45-47]. Yet, hypoparathyroidism usually spares the sacroiliac joint and this differentiates it from ankylosing spondylitis.

The treatment goals for AS focus on relieving pain and stiffness, maintaining axial spine mobility and functional ability, and preventing spinal complications. Non-pharmacological interventions should include regular exercise, postural training, and physical therapy.

The primary pharmacological approach involves the daily, long-term use of non-steroidal anti-inflammatory drugs (NSAIDs). If NSAIDs fail to provide adequate relief, they may be combined with or replaced by tumor necrosis factor inhibitors (TNF inhibitors) such as adalimumab, infliximab, or etanercept. The effectiveness of NSAIDs should be assessed 4 to 6 weeks after initiation, while the response to TNF inhibitors should be evaluated after 12 weeks.

Systemic glucocorticoids are generally not recommended for the treatment of AS; however, local steroid injections may be considered in specific cases. Referrals to specialists may be necessary depending on the patient's clinical presentation, potential complications, and extra-articular manifestations of the disease. Rheumatologists can assist with the formal diagnosis, management, and monitoring of AS, while dermatologists, ophthalmologists, and gastroenterologists may be involved in addressing associated non-musculoskeletal features of the condition.

Exercise programs have proven to be highly beneficial in managing ankylosing spondylitis (AS), particularly in addressing pain management, flexibility, mobility, and overall function. Although there is no universally recommended protocol for AS management, numerous studies have demonstrated improvements with various exercise approaches, including home exercise routines and group therapy sessions.

Hydrotherapy is frequently highlighted for its cardiovascular benefits and effectiveness in managing pain in individuals with AS. Additionally, respiratory and postural exercises are crucial, as the disease can adversely affect proper breathing mechanics and posture.

Patient education plays a key role in the management of AS. Patients should be informed about the chronic nature of the disease, the medications prescribed, and their potential side effects. Emphasis should be placed on the importance of regular exercise programs to alleviate symptoms. Physical therapy, including water therapy and swimming, can be particularly effective in reducing symptoms, enhancing functionality, and maintaining overall fitness.

Given the potential for pulmonary involvement in AS, smoking cessation should be strongly emphasized as part of the management strategy.

Conclusion

This case underscores the complexity of diagnosing idiopathic hypoparathyroidism (IHP), particularly when it presents with clinical features that can mimic other rheumatologic and skeletal disorders. The patient's presentation, marked by generalized seizures, clonic spasms, and an ankylosing spondylitis-like posture, initially led to a misdiagnosis of ankylosing spondylitis (AS). However, further investigation revealed a pattern of biochemical abnormalities—namely, hypocalcemia, hyperphosphatemia, and low parathyroid hormone (PTH) levels—that were consistent with hypoparathyroidism. The additional presence of cataracts and Fahr syndrome highlighted the long-standing nature of the patient's condition and the systemic impact of untreated hypoparathyroidism.

The diagnostic challenge in this case was compounded by the fact that the patient's skeletal changes, including exuberant intervertebral bone bridges and ossification at the superior acetabular margin, did not align with the typical radiologic findings of AS. These features are more consistent with the skeletal manifestations of chronic hypoparathyroidism, which include diffuse ligamentous and entheseal ossifications and spinal ligament calcification. The preservation of the sacroiliac joints further differentiated this case from AS, as sacroiliitis is a hallmark of the disease and a key diagnostic criterion.

The patient's eventual diagnosis of IHP was supported by the absence of a history of parathyroid surgery, radiation exposure, familial autoimmune polyendocrinopathies, and normal magnesium levels. This case highlights the importance of a comprehensive diagnostic approach, combining clinical assessment, biochemical testing, and imaging studies, to distinguish hypoparathyroidism from other similar conditions.

The effective management of IHP in this patient involved the initiation of calcium and vitamin D therapy, which led to significant clinical and biochemical improvements. The normalization of serum calcium levels, reduction in hyperphosphatemia, and improvement in the patient's neuromuscular and skeletal symptoms underscore the efficacy of this treatment approach. Additionally, the stabilization of the patient's condition and the resolution of symptoms such as seizures and irritability reflect the critical role of appropriate and timely intervention in improving patient outcomes.

This case serves as a valuable reminder of the need for clinicians to consider hypoparathyroidism in the differential diagnosis of patients presenting with chronic inflammatory symptoms and atypical skeletal findings. The systemic nature of hypoparathyroidism, with its potential impact on multiple organ systems and the development of chronic complications, necessitates a thorough evaluation and management strategy.

References

[1] Matikainen N, Pekkarinen T, Ryhänen EM, Schalin-Jäntti C. Physiology of Calcium Homeostasis: An Overview. *Endocrinol Metab Clin North Am.* 2021 Dec;50(4):575-590.

[2] Gafni RI, Collins MT. Hypoparathyroidism. *N Engl J Med.* 2019 May 2;380(18):1738-1747.

[3] Bilezikian JP. Hypoparathyroidism. *J Clin Endocrinol Metab.* 2020 Jun 1;105(6):1722-1736.

[4] Abate EG, Clarke BL. Review of Hypoparathyroidism. *Front Endocrinol (Lausanne).* 2016;7:172.

[5] Khan AA, Bilezikian JP, Brandi ML, Clarke BL, Gittoes NJ, Pasieka JL, et al. Evaluation and Management of Hypoparathyroidism Summary Statement and Guidelines from the Second International Workshop. *J Bone Miner Res.* 2022 Dec;37(12):2568-2585.

[6] Yao L, Guyatt G, Ye Z, Bilezikian JP, Brandi ML, Clarke BL, et al. Methodology for the Guidelines on Evaluation and Management of Hypoparathyroidism and Primary Hyperparathyroidism. *J Bone Miner Res.* 2022 Nov;37(11):2404-2410.

[7] Mannstadt M, Cianferotti L, Gafni RI, Giusti F, Kemp EH, Koch CA, et al. Hypoparathyroidism: Genetics and Diagnosis. *J Bone Miner Res.* 2022 Dec;37(12):2615-2629.

[8] Van Uum S, Shrayyef M, M'Hiri I, Dandurand K, Ali DS, Bilezikian JP, et al. Initial Assessment and Monitoring of Patients with Chronic Hypoparathyroidism: A Systematic Current Practice Survey. *J Bone Miner Res.* 2022 Dec;37(12):2630-2641.

[9] Yao L, Hui X, Li M, Li J, Ahmed MM, Lin C, et al. Complications, Symptoms, Presurgical Predictors in Patients With Chronic Hypoparathyroidism: A Systematic Review. *J Bone Miner Res. 2022* Dec;37(12):2642-2653.

[10] Rayannavar A, Levitt Katz LE, Crowley TB, Lessig M, Grand K, Goldmuntz E, et al. Association of hypocalcemia with congenital heart disease in 22q11.2 deletion syndrome. *Am J Med Genet A.* 2018 Oct;176(10):2099-2103.

[11] Cianferotti L, Marcucci G, Brandi ML. Causes and pathophysiology of hypoparathyroidism. *Best Pract Res Clin Endocrinol Metab.* 2018 Dec;32(6):909-925.

[12] Ehrenpreis ED, Jarrouj G, Meader R, Wagner C, Ellis M. A comprehensive review of hypomagnesemia. *Dis Mon.* 2022 Feb;68(2):101285.

[13] Bjornsdottir S, Ing S, Mitchell DM, Sikjaer T, Underbjerg L, Hassan-Smith Z, et al. Epidemiology and Financial Burden of Adult Chronic Hypoparathyroidism. *J Bone Miner Res.* 2022 Dec;37(12):2602-2614.

[14] Rao SD. Epidemiology of parathyroid disorders. *Best Pract Res Clin Endocrinol Metab.* 2018 Dec;32(6):773-780.

[15] Tinawi M. Disorders of Calcium Metabolism: Hypocalcemia and Hypercalcemia. *Cureus.* 2021 Jan 1;13(1).

[16] Bove-Fenderson E, Mannstadt M. Hypocalcemic disorders. *Best Pract Res Clin Endocrinol Metab.* 2018 Oct;32(5):639-656.

[17] Khan AA, Guyatt G, Ali DS, Bilezikian JP, Collins MT, Dandurand K, et al. Management of Hypoparathyroidism. *J Bone Miner Res.* 2022 Dec;37(12):2663-2677.

[18] Marcucci G, Cianferotti L, Brandi ML. Clinical presentation and management of hypoparathyroidism. *Best Pract Res Clin Endocrinol Metab.* 2018 Dec;32(6):927-939.

[19] Di Maio S, Soliman AT, De Sanctis V, Kattamis CC. Current treatment of hypoparathyroidism: Theory versus reality waiting guidelines for children and adolescents. *Acta Biomed.* 2018 Mar 27;89(1):122-131.

[20] Bollerslev J, Rejnmark L, Zahn A, Heck A, Appelman-Dijkstra NM, Cardoso L, et al. European Expert Consensus on Practical Management of Specific Aspects of Parathyroid Disorders in Adults and in Pregnancy: Recommendations of the ESE Educational Program of Parathyroid Disorders. *Eur J Endocrinol.* 2022 Feb 1;186(2).

[21] Gosmanova EO, Chen K, Rejnmark L, Mu F, Swallow E, Briggs A, et al. Risk of Chronic Kidney Disease and Estimated Glomerular Filtration Rate Decline in Patients with Chronic Hypoparathyroidism: A Retrospective Cohort Study. *Adv Ther.* 2021 Apr;38(4):1876-1888.

[22] Khan AA, Rubin MR, Schwarz P, Vokes T, Shoback DM, Gagnon C, et al. Efficacy and Safety of Parathyroid Hormone Replacement With TransCon PTH in Hypoparathyroidism: 26-Week Results From the Phase 3 PaTHway Trial. *J Bone Miner Res.* 2023 Jan;38(1):14-25.

[23] Yao L, Li J, Li M, Lin C, Hui X, Tamilselvan D, et al. Parathyroid Hormone Therapy for Managing Chronic Hypoparathyroidism: A Systematic Review and Meta-Analysis. *J Bone Miner Res.* 2022 Dec;37(12):2654-2662.

[24] Holmes EW, Samarasinghe S, Emanuele MA, Meah F. Biotin Interference in Clinical Immunoassays: A Cause for Concern. *Arch Pathol Lab Med.* 2017 Nov;141(11):1459-1460.

[25] Hakami Y, Khan A. Hypoparathyroidism. *Front Horm Res.* 2019;51:109-126.

[26] Linglart A, Levine MA, Jüppner H. Pseudohypoparathyroidism. *Endocrinol Metab Clin North Am.* 2018 Dec;47(4):865-888.

[27] Pepe J, Colangelo L, Biamonte F, Sonato C, Danese VC, Cecchetti V, et al. Diagnosis and management of hypocalcemia. *Endocrine.* 2020 Sep;69(3):485-495.

[28] Dershem R, Gorvin CM, Metpally RPR, Krishnamurthy S, Smelser DT, Hannan FM, et al. Familial Hypocalciuric Hypercalcemia Type 1 and Autosomal-Dominant Hypocalcemia Type 1: Prevalence in a Large Healthcare Population. *Am J Hum Genet.* 2020 Jun 4;106(6):734-747.

[29] Abate EG, Clarke BL. Review of Hypoparathyroidism. *Front Endocrinol (Lausanne)* 2017;7:172–172.

[30] Mannstadt M, Bilezikian JP, Thakker RV, Hannan FM, Clarke BL, Rejnmark L, et al. Hypoparathyroidism. *Nature Reviews Disease Primers* 2017;3:17055.

[31] Siraj N, Hakami Y, Khan A. Medical Hypoparathyroidism. *Endocrinology and Metabolism Clinics of North America* 2018;47:797–808.

[32] Mathieu S, Gossec L, Dougados M, Soubrier M. Cardiovascular profile in ankylosing spondylitis: A systematic review and meta-analysis. *Arthritis Care Res (Hoboken).* 2011 Apr;63(4):557-63.

[33] Zochling J, van der Heijde D, Burgos-Vargas R, Collantes E, Davis JC Jr, Dijkmans B, et al. ASAS/EULAR recommendations for the management of ankylosing spondylitis. *Ann Rheum Dis.* 2006 Apr;65(4):442-52.

[34] Rudwaleit M, van der Heijde D, Landewe R, Listing J, Akkoc N, Brandt J, et al. The development of Assessment of SpondyloArthritis international Society classification criteria for axial spondyloarthritis (part II): validation and final selection. *Ann Rheum Dis.* 2009 Jun;68(6):777-83.

[35] Dougados M, Baeten D. Spondyloarthritis. *Lancet.* 2011 Jun 18;377(9783):2127-37.

[36] Bennett AN, McGonagle D, O'Connor P, Hensor E, Sivera F, Coates LC, et al. Severity of baseline magnetic resonance imaging-evident sacroiliitis and HLA-B27 status in early inflammatory back pain predict radiographically evident ankylosing spondylitis at eight years. *Arthritis Rheum.* 2008 Dec;58(11):3413-8.

[37] Braun J, Sieper J. Ankylosing spondylitis. *Lancet.* 2007 Apr 21;369(9570):1379-90.

[38] van der Linden S, Valkenburg HA, Cats A. Evaluation of diagnostic criteria for ankylosing spondylitis. A proposal for modification of the New York criteria. *Arthritis Rheum.* 1984 Apr;27(4):361-8.

[39] Rao DS, Phillips ER, Divine GW, Talpos GB, Phillips RL. Randomized controlled clinical trial of nasal calcitonin in postmenopausal women with low bone mass. *Calcif Tissue Int.* 1993 Sep;53(3):209-15.

[40] Resnick D, Niwayama G. Radiographic and pathologic features of spinal involvement in diffuse idiopathic skeletal hyperostosis (DISH). *Radiology.* 1976 Nov;119(3):559-68.

[41] Goswami R, Ray D, Sharma R, Tomar N, Gupta R, Gupta N, et al. Presence of spondyloarthropathy and its clinical profile in patients with hypoparathyroidism. *Clinical Endocrinology* 2008;68:258–63.

[42] John DR, Suthar PP. Radiological Features of Long-Standing Hypoparathyroidism. *Pol J Radiol.* 2016;81:42-5

[43] Aynar R. Hip Pain and Gait Disturbance Associated with Idiopathic Hypoparathyroidism. *JCD* 2016.

[44] Ünverdi S, Öztürk MA, Inal S, Selek H, Göker B, Haznedaroglu çSeminur, et al. Idiopathic Hypoparathyroidism Mimicking Diffuse Idiopathic Skeletal Hyperostosis. *JCR: Journal of Clinical Rheumatology* 2009Oct;15(7):361-2.

[45] Iwai S, Kabata T, Maeda T, Kajino Y, Ogawa K, Kuroda K, et al. Hyperostosis around the bilateral acetabulum associated with hypoparathyroidism. Modern Rheumatology 2012;22:766–8.

[46] Theumann N, Aubry-Rozier B, Guntern D. Imagerie des atteintes inflammatoires rachidiennes. *Rev Med Suisse* 2009 ; 5: 566-70.

[47] Rudwaleit M, van der Heijde D, Landewé R, Listing J, Akkoc N, Brandt J, et al. The development of Assessment of SpondyloArthritis international Society classification criteria for axial spondyloarthritis (part II): validation and final selection. *Ann Rheum Dis* 2009;68:777.

Index

A

advanced technologies, 54, 85
alkaloid(s), 182, 186, 187, 193
analysis of variance (ANOVA), 3, 20, 42, 182
anatomical, 97, 98, 100, 107, 113, 115, 116, 117, 118, 119, 120
ankylosing spondylitis (AS), 195, 196, 197, 198, 199, 200, 201, 203, 204, 205, 206, 207, 208, 210, 211
anti-malarial, 121, 124, 125, 127, 140, 142
antimicrobial, 87, 99, 155, 157, 160, 165, 169, 170, 171, 172, 176, 192
avermectins (AVMs), 121, 125, 126, 127, 128, 138, 144, 145, 149

B

bignoniaceae, 176, 189, 192
bioactive metabolites, 176
biochemical, 4, 5, 55, 90, 170, 171, 175, 179, 186, 187, 189, 190, 193, 197, 201, 202, 207, 208
biological activities, 76, 99, 119, 145, 151, 152, 153, 160, 166, 168, 176
biomarker discovery, 54, 74, 76
biomedical, 54, 55, 75, 87, 94, 169
bioreactor(s), 167, 171, 173
biotechnology, 53, 91, 94, 151, 152, 153, 166, 168, 169, 170, 171, 173, 175, 177, 189, 190, 191
Bowman–Birk inhibitors, 151

C

cancer(s), 51, 72, 76, 85, 88, 89, 99, 115, 152, 153, 162, 164, 165, 173, 202
carbohydrate(s) (CB), 1, 2, 3, 4, 5, 13, 15, 16, 32, 34, 35, 36, 39, 43, 46, 193
catalase, 180, 185
cell suspension, 167, 171, 174
characterization, 46, 65, 75, 77, 90, 92, 94, 119, 145, 151, 152, 153, 157, 158, 161, 162, 169, 170, 171, 172, 174, 189, 193
chemometrics, 4, 6, 7, 8, 19, 21, 24, 37, 38, 39, 42, 44, 51
cross section(s), 98, 107, 108, 109, 110, 111, 112, 115, 116, 117
cryopreservation, 176, 178, 182, 183, 188, 189, 190, 191, 192
cryostorage, 177, 178

D

density functional theory (DFT), 3, 9, 10, 11, 12, 22, 28, 29, 30, 49, 51
dosage(s), 122, 130, 132, 134, 135, 136, 139
dosing, 127, 137, 139, 140, 142
drug development, 54, 153

E

electrospray ionization (ESI), 3, 47, 55, 64, 88, 129
electrostatic potentials (ESP), 3, 11, 28, 29, 30, 49
Ex situ conservation, 175, 177, 186

extraction(s), 41, 57, 60, 85, 86, 87, 88, 94, 129, 155, 157, 158, 159, 160, 161, 162, 163, 164, 165, 166, 179

F

Fabaceae (Leguminosae), 152, 157, 158, 159, 161, 162, 163, 172
flavonoid(s), 176, 181, 186, 187, 190, 192
flor-de-São-João, 176

G

gas chromatography and mass spectrometry (GC–MS), 1, 3, 4, 13, 18, 20, 32, 34, 35, 36, 38, 39, 41, 43
germination, 152, 159, 176, 177, 178, 179, 183, 184, 187, 189, 190
glutathione, 180, 185, 190
ground state (GS), 3, 9, 21, 22, 24, 25, 26, 27, 29, 31, 32, 40, 42
growth, 4, 52, 73, 77, 127, 138, 155, 158, 160, 163, 166, 167, 179, 183, 186, 187, 188, 189, 191

H

human host, 122, 138, 139
hybrid(s), 29, 33, 91, 96, 140, 141, 143
hypocalcemia, 196, 197, 199, 202, 203, 207, 209, 210
hypoparathyroidism (HP), 3, 21, 22, 25, 26, 27, 29, 31, 32, 40, 42, 195, 196, 197, 200, 202, 203, 206, 207, 208, 209, 210, 211

I

identification key, 98, 100, 118
idiopathic hypoparathyroidism (IHP), 195, 196, 197, 199, 200, 202, 203, 207, 208, 211
In vitro cultivation, 176, 177, 178, 183, 186
In vitro germination, 175, 178, 189
In vitro production, 151, 153, 155, 157
infection(s), 76, 80, 88, 115, 122, 123, 124, 125, 126, 127, 131, 132, 134, 138, 140, 142, 147, 149, 151, 169
inflammation, 99, 115, 151, 153, 155, 161, 165, 173, 198, 204, 205
ivermectin (IVM), 121, 122, 125, 126, 127, 129, 130, 132, 133, 134, 135, 136, 137, 138, 139, 140, 142, 144, 145, 146, 147, 148, 149, 150

K

Kunitz, 151, 152, 153, 154, 164, 168, 169, 170, 171, 172, 173, 174
Kunitz-type protease inhibitors (KTPIs), 151, 152, 153, 154, 155, 157, 161, 162, 165, 166, 167, 168

L

leaf, 13, 40, 41, 43, 98, 100, 105, 111, 112, 113, 115, 116, 119, 120
leaves, 1, 3, 4, 13, 14, 15, 16, 41, 43, 46, 95, 98, 99, 100, 101, 103, 105, 114, 115, 116, 118, 191
linoleic acid, 1, 2, 3, 4, 13, 15, 16, 17, 18, 19, 20, 21, 22, 25, 27, 31, 43, 46
lipoperoxidation (LP), 180, 185, 186
liquid nitrogen, 175, 177, 179, 184, 185, 187, 189
low molecular weight (LMW), 3, 6, 51

M

macrocyclic, 121, 125, 127, 145
malaria, 121, 122, 123, 124, 125, 127, 130, 131, 132, 133, 134, 135, 136, 137, 138, 139, 140, 143, 144, 145, 146, 147, 148, 149
mass spectrometry (MS), 1, 3, 4, 5, 6, 7, 8, 10, 13, 17, 18, 19, 20, 21, 28, 29, 30, 32, 34, 35, 36, 38, 39, 41, 43, 46, 47, 51, 53, 54, 55, 56, 57, 58, 59, 60, 61, 62, 63, 64, 65, 67, 68, 69, 70, 71, 73, 74, 75, 76, 77, 78, 79, 80, 81, 82, 83, 84, 85, 86, 87, 88,

89, 90, 91, 92, 93, 94, 95, 96, 129, 160, 179
mechanism(s), 4, 10, 46, 54, 69, 76, 77, 85, 124, 130, 139, 143, 151, 164, 169, 170, 171, 186, 188, 197
medicinal plant(s), 97, 99, 100, 102, 115, 119, 176, 189
metabolomic(s), 2, 3, 5, 6, 14, 17, 18, 21, 34, 44, 45, 46, 47, 52, 85, 86, 119
moisture, 176, 178, 182, 189
molecular dynamics (MD), 3, 4, 9, 10, 21, 22, 23, 49
molecular mechanics (MM), 4, 10, 48, 49, 209
morphological, 97, 98, 99, 100, 102, 113, 116, 118, 120, 183

N

natural bond orbital (NBO), 3, 4, 11, 29, 33, 49

O

oxidative parameters, 176, 179

P

parasite(s), 122, 123, 124, 125, 126, 127, 128, 130, 132, 133, 134, 135, 136, 138, 139, 140, 169
parathyroid hormone (PTH), 195, 196, 197, 199, 200, 202, 203, 207, 210
petiole, 41, 98, 103, 111, 112, 116, 117
phenolic compound(s), 4, 176, 181, 186, 187, 189
phenotypic characteristic(s), 176, 183
phenylalanine ammonia-lyase (PAL), 181, 185, 186, 191
phytochemical, 100, 119, 181, 191
phytosterols, 182, 186, 187
plant(s), 1, 4, 7, 13, 17, 40, 41, 43, 44, 45, 46, 52, 95, 97, 99, 100, 101, 102, 107, 116, 117, 118, 119, 120, 151, 152, 153, 154, 155, 157, 158, 159, 161, 162, 164, 166, 167, 168, 169, 170, 171, 172, 173, 174, 175, 176, 177, 178, 183, 186, 188, 189, 190, 191,192, 193
Plantago, 97, 98, 99, 100, 102, 103, 108, 109, 110, 115, 116, 117, 118, 119, 120
plasmodium, 122, 124, 130, 134, 138, 140, 143, 145, 146, 147, 149
polarizable continuum method (PCM), 4, 9
potential energy surface (PES), 4, 9, 10, 12
protease inhibitor(s) (PIs), 151, 152, 154, 160, 161, 164, 166, 167, 168, 169, 170, 171, 172, 173, 174
proteases, 56, 58, 62, 151, 152, 154, 169, 170, 171, 173
proteomic(s), 46, 47, 52, 53, 54, 55, 56, 57, 61, 64, 65, 67, 68, 69, 70, 71, 74, 75, 77, 79, 80, 81, 82, 83, 84, 85, 86, 87, 88, 89, 90, 91, 92, 93, 94, 95, 96, 170
purification, 57, 61, 79, 88, 96, 151, 152, 153, 157, 159, 161, 162, 163, 164, 166, 168, 169, 170, 171
Pyrostegia venusta, 175, 176, 178, 179, 184, 185, 187, 189, 190, 191, 192, 193

Q

quantum chemistry, 3

R

radiological, 195, 198, 202, 211
reducing sugars (RS), 180, 185, 186
resistance, 43, 49, 52, 128, 135, 139, 140, 143, 149, 150, 151
retention time (RT), 4, 14, 15, 34, 37, 49, 66, 74, 129
root(s), 45, 88, 100, 101, 107, 116, 117, 176, 179, 181, 183, 184, 185, 186, 187, 188, 192

S

scape, 98, 100, 108, 109, 110, 111, 117, 118

secondary metabolites, 1, 4, 175, 176, 177, 186, 187, 191
seed(s), 98, 99, 100, 104, 105, 107, 116, 118, 119, 120, 151, 152, 153, 155, 157, 158, 159, 160, 161, 162, 163, 164, 165, 168, 169, 170, 171, 172, 174, 175, 177, 178, 179, 182, 183, 184, 185, 186, 187, 189, 190, 191, 192, 193
short-term liquid nitrogen, 175, 179, 189
Solanum lycopersicum L., 1, 3, 4, 13
soluble proteins, 179
soluble sugars (SS), 180, 185, 186, 192
species, 8, 9, 10, 11, 13, 21, 22, 28, 29, 30, 31, 33, 40, 54, 97, 98, 99, 100, 102, 105, 115, 118, 119, 120, 122, 131, 132, 134, 135, 137, 140, 147, 152, 157, 158, 160, 161, 162, 170, 175, 176, 177, 178, 186, 187, 188, 190, 191
sporogony, 131, 134, 135, 136, 146
stochastic dynamics, 2, 3, 6, 7, 13, 17, 18, 21, 31, 43, 47
storage(s), 46, 154, 167, 169, 175, 177, 179, 182, 184, 187, 189
structural analysis, 2, 3, 5, 6, 7, 13, 17, 18, 21, 28, 44, 47
superficial, 98, 114, 115, 116

T

tomato leaves, 1, 4, 13, 14, 15, 16, 41, 43
transition state (TS), 4, 8, 9, 22, 23, 25, 26, 27, 29, 31, 32, 33, 40

V

validation, 65, 68, 69, 71, 81, 83, 87, 129, 210, 211
vector(s), 122, 125, 126, 127, 130, 131, 132, 133, 135, 136, 137, 144, 145, 146, 147, 148, 149

W

wonder drug, 122, 128, 142, 145